The Engineering Dynamics Course Companion, Part 2: Rigid Bodies

Kinematics and Kinetics

Complete Supplemental Video Playlist

https://www.youtube.com/playlist?list=PL5aZISlMu3kJnmZ7VX8w3bOLH0SBGg2xn

Synthesis Lectures on Mechanical Engineering

Synthesis Lectures on Mechanical Engineering series publishes 60–150 page publications pertaining to this diverse discipline of mechanical engineering. The series presents Lectures written for an audience of researchers, industry engineers, undergraduate and graduate students.

Additional Synthesis series will be developed covering key areas within mechanical engineering.

The Engineering Dynamics Course Companion, Part 2: Rigid Bodies: Kinematics and Kinetics
Edward Diehl
2020

The Engineering Dynamics Course Companion, Part 1: Particles: Kinematics and Kinetics
Edward Diehl
2020

Introduction to Deep Learning for Engineers: Using Python on Google Cloud Platform
Tariq M. Arif
2020

Towards Analytical Chaotic Evolutions in Brusselators
Albert C.J. Luo and Siyu Guo
2020

Modeling and Simulation of Nanofluid Flow Problems
Snehashi Chakraverty and Uddhaba Biswal
2020

Modeling and Simulation of Mechatronic Systems using Simscape
Shuvra Das
2020

Automatic Flight Control Systems
Mohammad Sadraey
2020

The Engineering Dynamics Course Companion, Part 2: Rigid Bodies: Kinematics and Kinetics
Edward Diehl

ISBN: 978-3-031-79680-7   paperback
ISBN: 978-3-031-79681-4   ebook
ISBN: 978-3-031-79682-1   hardcover

DOI 10.1007/978-3-031-79681-4

A Publication in the Springer series
SYNTHESIS LECTURES ON MECHANICAL ENGINEERING

Lecture #26
Series ISSN
Print 2573-3168   Electronic 2573-3176

The Engineering Dynamics Course Companion, Part 2: Rigid Bodies

Kinematics and Kinetics

Edward Diehl
University of Hartford

SYNTHESIS LECTURES ON MECHANICAL ENGINEERING #26

ABSTRACT

Engineering Dynamics Course Companion, Part 2: Rigid Bodies: Kinematics and Kinetics is a supplemental textbook intended to assist students, especially visual learners, in their approach to Sophomore-level Engineering Dynamics. This text covers particle kinematics and kinetics and emphasizes Newtonian Mechanics "Problem Solving Skills" in an accessible and fun format, organized to coincide with the first half of a semester schedule many instructors choose, and supplied with numerous example problems. While this book addresses Rigid Body Dynamics, a separate book (Part 1) is available that covers Particle Dynamics.

KEYWORDS

dynamics, particle kinematics, particle kinetics, Newtonian mechanics

Contents

Acknowledgments

This course companion is the result of a decade of teaching Dynamics in close cooperation with several brilliant and dedicated engineering educators. I would like to sincerely thank my colleagues and mentors for their assistance and inspiration. Here I acknowledge their contribution to this effort and my career as an educator in reverse chronological order.

I'm grateful to my fellow faculty at the University of Hartford, many of whom reviewed the manuscript and offered extremely valuable and insightful feedback and corrections. These include Dr. Cy Yavuzturk, Dr. Mark Orelup, Dr. Mary Arico, Dr. Taka Asaki, Professor Phil Faraci, and Dr. Chris Jasinski. I'm indebted to my Ph.D. advisor, friend, and mentor, Dr. Jiong Tang, for his support and encouragement to publish a work that mattered to me. I'm forever thankful to my friends and colleagues at the United States Coast Guard Academy for supporting me during and providing the opportunity to transition from a practicing engineer to an educator. These include Dr. Todd Taylor, Captain Mike Corl, Dr. Elisha Garcia, Lieutenant (Ret) Sean Munnis, Commander Nick Parker, Dr. Tom DeNucci, Dr. Susan Swithenbank, Commander John Goshorn, and Lieutenant Commander J.J. Schock. The close working relationship of these instructors in which we shared notes, examples, and exam problems heavily influenced the content of this book, and many of the problems within are adaptations of this group effort. I'd like to acknowledge Sean Munnis in particular as the person who dubbed Sir Isaac Newton "Newtdog" and encouraged me more than anyone to draw him as a cartoon character and develop this into a book. Before I joined academia, I was a working engineer and I'm grateful for my former colleagues at Seaworthy Systems and General Dynamics, but especially my mentor, the late Bill McCarthy, who pushed me and inspired me to be a better engineer and better writer. Special thanks to the late Professor Don Paquette of the United States Merchant Marine Academy, my Statics, Dynamics, and Machine Design professor. He inspired me to become a teacher, and I've endeavored to follow in his footsteps.

And in real life, I'm so very thankful to my incredibly supportive wife, Lori Dappert Diehl, who actually read this book. Thank you, Lori, for always making me laugh and never letting me give up. Lastly, I'd like to acknowledge my older brother, the late James Harold Diehl, whose communication limitations have obliged me to communicate and whose resilience inspires me to persevere through the relatively minor inconveniences of life.

Edward Diehl
August 2020

Book 1 - Class 0

https://www.youtube.com/watch?v=WvowYa_3OQg

CLASS   0

Introduction

B.L.U.F. (Bottom Line Up Front)

- Dynamics is the study of motion.

- Kinematics and Kinetics:

    - Kinematics: the description of motion, ignoring the cause of the motion.

    - Kinetics: the interaction of loading and motion on objects with mass.

- Categories of objects:

    - Particles: objects treated as point masses since their size and shape isn't important.

    - Rigid Bodies: objects whose size and shape m their rotation is important to how they move.

- Course is broken down into four parts.

    - Particle Kinematics, Particle Kinetics, Rigid Body Kinematics, Rigid Body Kinetics.

0.1   ABOUT THE BOOK

This is the second part of a two-part "course companion" to assist undergraduate engineering students taking a first course in Dynamics. Part 1 deals with the dynamics of particles, while Part 2 covers the dynamics of rigid bodies. Your course companions are "Newtdog and Wormy" (Figure 1) who will guide you through Newtonian Dynamics with plenty of examples and images especially geared towards visual learners. Much of the content from the introduction to Part 1 is repeated in this introduction so Part 2 is equally useful on its own.

Many engineering majors typically take Dynamics in their Sophomore year after completing Statics, and this is one of the most challenging transitions, requiring considerable personal growth in the way problems are approached and processed. This book is meant to help with that transition and serve as a "course companion," a complementary resource a struggling student can refer to when frustrated.

Figure 1: Portrait of Sir Isaac Newton by Godfrey Kneller with your course companions: Newt-dog and Wormy (© E. Diehl).

Why is Dynamics so difficult? Many of the problem types require students to think differently than they're used to: rely less on step-by-step procedures and instead recognize the nature of a problem and navigate to a solution using concepts. Sometimes problems require working backward or applying logic to generate an "ah-ha" moment, when the lightbulb goes off and the path to a solution becomes clear. Students often describe some Dynamics assignments as "trick problems." This is true in a way: the solution will seem obvious once revealed. A good problem solver doesn't need to have worked through an identical problem in order to solve a new problem they've never seen. Instead, with problem solving experience, they develop a skill to pick it apart, identify the underlying principles, and formulate a path forward. Sometimes this is like a maze, where going down one path leads to a dead end. Problem solvers know to reverse course a bit, revaluate, and try a new approach. This text is intended to be your companion on that journey to developing "ah-ha" skills.

Because this is a "course companion," the book is written in a relatively casual tone compared to most textbooks (note the frequency of the pronoun "we") and includes Sir Isaac Newton as a cartoon to add some levity to this often-dreaded course. The cartoons are intended to also serve as "visual mnemonics." That is, they are meant to be memorable with an aspect of them associated with particular concepts as they're presented. Solving problems in dynamics requires recognizing the nature of a problem, identifying the key concepts, and applying a solution strategy. The middle part is where these cartoons can help, especially if one can think "oh, this is just like when _____." The blank being an aspect of the cartoon.

The 2 parts of this course companion consist of 12 "classes," each coinciding with the typical 2-class-a-week schedule of a semester-long Dynamics course. A common complaint of Dynamics students is not having enough examples or that the available examples are much easier than the homework. Therefore, the examples within each class are progressively longer and more challenging. Some textbooks skip steps within the example solutions, so this course companion attempts to work through the solutions in exhaustive detail. Example exam questions are also included in the appendices to provide opportunity for additional problem-solving practice. This course companion is also intended to assist instructors seeking inspiration for their own examples, homework problems and exam problems.

0.2   NEWTDOG AND WORMY: YOUR COURSE COMPANIONS

"Newtdog" is a silly nickname for Sir Isaac Newton intended to make him less intimidating. Based on quotes, his own writings, and biographies, Sir Isaac Newton seems to have been a down-to-Earth regular guy who was inquisitive and humble. He said: "If I have seen further than others, it is by standing upon the shoulders of giants. " and "To myself I am only a child playing on the beach, while vast oceans of truth lie undiscovered before me."

Newtdog is drawn to seem friendly, adventurous (just as Sir Isaac Newton was revolutionary), and a little bit of a dandy with his powered wig, frilly cuffs, long coat, and buckled shoes. Newtdog's buddy is "Wormy" who lives in the iconic apple that apocryphally led Newton to "discover gravity." Wormy is often just along for the ride and a little nervous about Newtdog's enthusiasm and adventurous spirit.

0.3   BOTTOM LINE UP FRONT (B.L.U.F.)

Every chapter begins with a "Bottom Line Up Front" (B.L.U.F.) consisting of bulleted items of the contents with very brief summaries and/or equations. The purpose is to introduce you to the essentials of the topic(s) covered and serve as a quick reference for later use when flipping through the book to search for content. In a classroom environment the BLUF provided at the beginning of class helps get the students prepared for what they're about to learn. Students using this course companion should read the BLUF just before class (at a minimum) so you're on the lookout for this information. The BLUF only takes a few seconds, so it's easy.

0.4   KINEMATICS VS. KINETICS

Dynamics can be organized into two parts: Kinematics and Kinetics. It's useful to memorize the definition of these terms to help organize the approaches we'll take.

Kinematics is the description of motion without regards to why it's happening and studies the relationships among time, position, velocity, and acceleration of an object. Kinematics is also often described as "the geometry of motion." We'll begin with particle kinematics since it is perhaps the simplest Dynamics broad topic and introduces many fundamental sub-topics which are useful to build upon.

Kinetics investigates why motion occurs and the interaction between loads and mass. The approach taken here falls into the category of Newtonian Mechanics since it's based on the principals described by Sir Isaac Newton in 1687. This is considered Classical Mechanics which also includes Lagrangian (1788) and Hamiltonian (1833) Mechanics that are reformulations of Newton's approach. You've likely also heard of Quantum Mechanics which shows that Classical Mechanics breaks down on the atomic and sub-atomic level. There are many other methods of studying motion, but Newtonian remains the cornerstone of an engineering education.

Another categorization of Dynamics topics is by Particles and Rigid Bodies. Particles, covered in Part 1, are point masses whose shapes aren't considered important enough to be included in the analysis. Rigid Bodies, covered in Part 2, have a shape significant enough to include the effect of rotation into the analysis, but their flexibility isn't sufficient to influence the results.

0.5   COURSE BREAKDOWN

The course companion (Parts 1 and 2) are organized to follow the classes of a typical Dynamics course. Part 1 covers Particle Kinematics and Kinetics, and Part 2 covers Rigid Body Kinematics and Kinetics.

- Part 1:
    - Kinematics of Particles
        1. Rectilinear Motion of Particles
        2. Special Cases and Relative Motion
        3. Curvilinear Motion of Particles and Projectile Motion (Rectangular)
        4. Non-Rectangular Coordinates (Path)
        5. Non-Rectangular Coordinates (Polar)
    - Kinetics of Particles
        6. Newton's Second Law in Rectangular Coordinates
        7. Newton's Second Law in Path and Polar Coordinates
        8. Work and Energy, Conservation of Energy
        9. Work and Energy, Conservation of Energy (Part 2)
        10. Impulse and Momentum
        11. Direct Central Impact

12. Oblique Central Impact

- Part 2:

  – Kinematics of Rigid Bodies

    13. Translation and Fixed Axis Rotation
    14. General Plane Motion, Absolute and Relative Velocity
    15. Instantaneous Center of Rotation
    16. General Plane Motion: Acceleration
    17. General Plane Motion: Acceleration (Part 2)
    18. Analyzing Motion w.r.t. a Rotating Frame (Coriolis)

  – Kinetics of Rigid Bodies

    19. Mass Moment of Inertia
    20. Newton's Second Law in Constrained Plane Motion
    21. Newton's Second Law in Translation and Rotation Plane Motion
    22. Energy Methods
    23. Momentum Methods
    24. Eccentric Impact

The topics are broken down in this manner to coincide with a 2 class per week, 14-week semester arrangement. Given these 28 possible class periods and subtracting 3 periods for exams the last day of class for review, 24 classes are available to introduce topics. Note there are two classes (9 and 17) that repeat the previous class topic. These are included to give more emphasis to those topics that might otherwise be too much information to absorb in one class. Experience has shown this to be a practical schedule for this level of an Engineering Dynamics course. Table 0.1 presents a suggested course schedule.

0.6   EQUATION SHEET

Table 0.2 presents a suggested equation sheet for instructors who choose to provide one during exams rather than have students make their own. It is purposefully limited to only two sheets of equations and does not include every permutation of the equations but enough to avoid students' having to memorize formulae. Students who have the option to write their own equation sheet should refer to this to ensure they've covered all the essentials.

Table 0.1: Course schedule

Week	Class	Topic
1	1	Kinematics of Particles - Rectilinear Motion of Particles
1	2	Kinematics of Particles - Special Cases: Relative and Dependent Motion
2	3	Kinematics of Particles - Curvilinear Motion of Particles (Rectangular)
2	4	Kinematics of Particles - Non-Rectangular Components (Path)
3	5	Kinematics of Particles - Non-Rectangular Components (Path)
3		**Exam 1 (Covering Classes 1–5)**
4	6	Kinetics of Particles - Newton's Second Law in Rectangular Coordinates
4	7	Kinetics of Particles - Newton's Second Law in Path and Polar Coordinates
5	8	Kinetics of Particles - Work and Energy and the Conservation of Energy (Part 1)
5	9	Kinetics of Particles - Work and Energy and the Conservation of Energy (Part 2)
6	10	Kinetics of Particles - Impulse-Momentum Method
6	11	Kinetics of Particles - Direct Impact of Particles and the Conservation of Linear Momentum
7	12	Kinetics of Particles - Oblique Impact of Particles
7		**Exam 2 (Covering Classes 8–12)**
8	13*	Kinematics of Rigid Bodies - Angular Kinematics of Rigid Body Motion
8	14	Kinematics of Rigid Bodies - Absolute and Relative Velocity
9	15	Kinematics of Rigid Bodies - Velocity Analysis Using the Instantaneous Center of Rotation
9	16	Kinematics of Rigid Bodies - Acceleration Analysis (Part 1)
10	17	Kinematics of Rigid Bodies - Acceleration Analysis (Part 2)
10	18	Kinematics of Rigid Bodies - Coriolis Acceleration Analysis
11		**Exam 3 (Covering Classes 13–18)**
11	19	Kinetics of Rigid Bodies - Mass Moment of Inertia
	20	Kinetics of Rigid Bodies - Newton's Second Law in Constrained Plane Motion
	21	Kinetics of Rigid Bodies - Newton's Second Law in Translating and Rotating Plane Motion
13	22	Kinetics of Rigid Bodies - Rigid Body Work-Energy Method
13	23	Kinetics of Rigid Bodies - Rigid Body Impulse-Momentum Method
14	24	Kinetics of Rigid Bodies - Impact of Rigid Bodies
14	25	Course Summary and Review for Final Exam
		Final Exam (Covering Entire Semester but emphasizing Classes 19–24)

* Classes 1–12 are covered in *Engineering Dynamics Course Companion, Part 1: Particles*

Table 0.2: Dynamics exam equation sheet (*Continues.*)

Particle Kinematics	Rigid Body Kinematics

Particle Kinematics

Velocity and acceleration in rectilinear motion:

$$\vec{\mathbf{v}} = \frac{d\vec{\mathbf{r}}}{dt} \qquad \vec{\mathbf{a}} = \frac{d\vec{\mathbf{v}}}{dt} = \frac{d^2\vec{\mathbf{r}}}{dt^2} = \vec{\mathbf{v}}\frac{d\vec{\mathbf{v}}}{d\vec{\mathbf{r}}}$$

Uniform translational motion:

$$\vec{\mathbf{r}} = \vec{\mathbf{r}}_0 + \vec{\mathbf{v}}_c t \qquad\qquad x = x_0 + v_{x,c}t$$

Uniformly accelerated translational motion:

$$\vec{\mathbf{v}} = \vec{\mathbf{v}}_0 + \vec{\mathbf{a}}_c t \qquad\qquad v_x = (v_0)_x + a_{x,c}t$$

$$\vec{\mathbf{r}} = \vec{\mathbf{r}}_0 + \vec{\mathbf{v}}_0 t + \tfrac{1}{2}\vec{\mathbf{a}}_c t^2 \qquad x = x_0 + (v_0)_x t + \tfrac{1}{2}a_{x,c}t^2$$

$$\vec{\mathbf{v}}^2 = \vec{\mathbf{v}}_0^2 + 2\vec{\mathbf{a}}_c(\vec{\mathbf{r}}-\vec{\mathbf{r}}_0) \qquad v_x^2 = (v_0)_x^2 + a_{x,c}(x-x_0)$$

Relative motion of two particles (or points):

$$\vec{\mathbf{r}}_B = \vec{\mathbf{r}}_A + \vec{\mathbf{r}}_{B/A} \qquad\qquad x_B = x_A + x_{B/A}$$

$$\vec{\mathbf{v}}_B = \vec{\mathbf{v}}_A + \vec{\mathbf{v}}_{B/A} \qquad\qquad v_B = v_A + v_{B/A}$$

$$\vec{\mathbf{a}}_B = \vec{\mathbf{a}}_A + \vec{\mathbf{a}}_{B/A} \qquad\qquad a_B = a_A + a_{B/A}$$

Path (tangential and normal) components:

$$\vec{\mathbf{v}} = v_t\hat{\mathbf{e}}_t = (v)\hat{\mathbf{e}}_t$$

$$\vec{\mathbf{a}} = a_t\hat{\mathbf{e}}_t + a_n\hat{\mathbf{e}}_n = \frac{dv}{dt}\hat{\mathbf{e}}_t + \frac{v^2}{\rho}\hat{\mathbf{e}}_n$$

Polar (radial and transverse) components:

$$\vec{\mathbf{v}} = v_r\,\hat{\mathbf{e}}_r + v_\theta\,\hat{\mathbf{e}}_\theta = (\dot{r})\hat{\mathbf{e}}_r + (r\dot{\theta})\hat{\mathbf{e}}_\theta$$

$$\vec{\mathbf{a}} = a_r\,\hat{\mathbf{e}}_r + a_\theta\,\hat{\mathbf{e}}_\theta = (\ddot{r} - r\dot{\theta}^2)\hat{\mathbf{e}}_r + (r\ddot{\theta} + 2\dot{r}\dot{\theta})\hat{\mathbf{e}}_\theta$$

Rigid Body Kinematics

Rotation about a fixed axis:

$$\vec{\mathbf{v}} = \frac{d\vec{\mathbf{r}}}{dt} = \vec{\boldsymbol{\omega}}\times\vec{\mathbf{r}}, \qquad \vec{\boldsymbol{\omega}} = \omega\hat{\mathbf{k}} = \dot{\theta}\hat{\mathbf{k}}$$

$$\vec{\mathbf{a}} = \underbrace{\vec{\boldsymbol{\alpha}}\times\vec{\mathbf{r}}}_{\vec{\mathbf{a}}_t} + \underbrace{\vec{\boldsymbol{\omega}}\times(\vec{\boldsymbol{\omega}}\times\vec{\mathbf{r}})}_{\vec{\mathbf{a}}_n}, \qquad \vec{\boldsymbol{\alpha}} = \alpha\hat{\mathbf{k}} = \ddot{\theta}\hat{\mathbf{k}}$$

$$a_t = r\alpha, \quad a_n = r\omega^2 \text{ (in one plane)}$$

Angular velocity and angular acceleration:

$$\vec{\boldsymbol{\omega}} = \frac{d\vec{\boldsymbol{\theta}}}{dt} \qquad\qquad \vec{\boldsymbol{\alpha}} = \frac{d\vec{\boldsymbol{\omega}}}{dt} = \frac{d^2\vec{\boldsymbol{\theta}}}{dt} = \vec{\boldsymbol{\omega}}\frac{d\vec{\boldsymbol{\omega}}}{d\vec{\boldsymbol{\theta}}}$$

Uniform rotational motion:

$$\theta = \theta_0 + \omega_c t$$

Uniformly accelerated rotational motion:

$$\omega = \omega_0 + \alpha_c t$$

$$\theta = \theta_0 + \omega_0 t + \tfrac{1}{2}\alpha_c t^2$$

$$\omega^2 = \omega_0^2 + 2\alpha_c(\theta - \theta_0)$$

Velocity in plane motion:

$$\vec{\mathbf{v}}_B = \vec{\mathbf{v}}_A + \vec{\mathbf{v}}_{B/A} = \vec{\mathbf{v}}_A + \omega\hat{\mathbf{k}}\times\vec{\mathbf{r}}_{B/A}$$

Acceleration in plane motion:

$$\vec{\mathbf{a}}_B = \vec{\mathbf{a}}_A + \vec{\mathbf{a}}_{B/A} = \vec{\mathbf{a}}_A + (\vec{\mathbf{a}}_{B/A})_t + (\vec{\mathbf{a}}_{B/A})_n$$

$$= \vec{\mathbf{a}}_A + \alpha\hat{\mathbf{k}}\times\vec{\mathbf{r}}_{B/A} - \omega^2\,\vec{\mathbf{r}}_{B/A}$$

Relative Motion <u>on</u> a Rigid Body:

Velocity of a point on a rigid body in plane motion:

$$\vec{\mathbf{v}}_B = \vec{\mathbf{v}}_A + \omega\hat{\mathbf{k}}\times\vec{\mathbf{r}}_{B/A} + \vec{\mathbf{v}}_{rel}$$

Acceleration of a point on a rigid body in plane motion:

$$\vec{\mathbf{a}}_B = \vec{\mathbf{a}}_A + \alpha\hat{\mathbf{k}}\times\vec{\mathbf{r}}_{B/A} - \omega^2\,\vec{\mathbf{r}}_{B/A} + \vec{\mathbf{a}}_{rel} + 2\vec{\boldsymbol{\omega}}\times\vec{\mathbf{v}}_{rel}$$

Table 0.2: (*Continued.*) Dynamics exam equation sheet

Particle Kinetics

Linear momentum of a particle: $\vec{L} = m\vec{v}$

Angular momentum of a particle: $\vec{H} = \vec{r} \times m\vec{v}$

Newton's second law: $\Sigma \vec{F} = m\vec{a} = \dot{\vec{L}}$

Equations of motion for a particle:

Cartesian: $\Sigma F_x = ma_x$ $\Sigma F_y = ma_y$ $\Sigma F_z = ma_z$

Path coord.: $\Sigma F_t = m\dfrac{dv}{dt}$ $\Sigma F_n = m\dfrac{v^2}{\rho}$

Polar coord: $\Sigma F_r = m(\ddot{r} - r\dot{\theta}^2)$ $\Sigma F_\theta = m(r\ddot{\theta} + 2\dot{r}\dot{\theta})$

Principle of work and energy:

$$KE_1 + PE_1 + U_{1\to2} = KE_2 + PE_2$$

Work of a force: $U_{1\to2} = \int_{A_1}^{A_2} \vec{F} \cdot d\vec{r}$

Kinetic energy of a particle: $KE = \frac{1}{2} mv^2$

Potential energy: $PE_g = mgy, PE_{sp} = \frac{1}{2} kx^2$

Power and mechanical efficiency:

Power $= \dfrac{dW}{dt} = \dot{U} = \vec{F} \cdot \vec{v}$ or $= \dfrac{Md\theta}{dt} = M\omega$

$\eta = \dfrac{\text{energy (or power) output}}{\text{energy (or power) input}}$

Principle of impulse and momentum for particles:

$$\vec{L}_1 + \overrightarrow{\textbf{IMP}}_{1\to2} = \vec{L}_2$$

$$\Sigma m\vec{v}_1 + \int_{t_1}^{t_2} \vec{F}\, dt = \Sigma m\vec{v}_2$$

Oblique central impact:

$$(v_A)_t = (v_A')_t, \qquad (v_B)_t = (v_B')_t$$

$$m_A(v_A)_n + m_B(v_B)_n = m_A(v_A')_n + m_B(v_B')_n$$

$$(v_B')_n - (v_A')_n = e[(v_A)_n - (v_B)_n]$$

Rigid Body Kinetics

Mass center: $m\vec{r} = \sum_{i=1}^{n} m_i \vec{r}_i$

Moments of inertia of masses and radius of gyration:

$$I = \int r^2\, dm \qquad k = \sqrt{\dfrac{I}{m}}$$

Parallel-axis theorem: $I = \bar{I} + md^2$

Equations for the plane motion of a rigid body:

$$\Sigma F_x = m\bar{a}_x \qquad \Sigma F_y = m\bar{a}_y$$

$$\Sigma M_O = \Sigma(M_O)_{eff} = \bar{I}\alpha + ma_G\, r_{G/O}$$

Work of a couple of moment M:

$$U_{1\to2} = \int_{\theta_1}^{\theta_2} \vec{M} \cdot d\vec{\theta}$$

Kinetic energy in plane motion:

$$KE = \frac{1}{2} m\bar{v}^2 + \frac{1}{2}\bar{I}\omega^2 = \frac{1}{2} I_O \omega^2$$

Principle of impulse and momentum for a rigid body:

Syst Momenta₁ + Syst Ext Imp₁→₂ = Syst Momenta₂

Angular momentum in plane motion about mass center:

$$\vec{H}_G = \bar{I}\vec{\omega}$$

Principle of impulse and momentum for particles:

$$(\vec{H}_O)_1 + (\overrightarrow{\textbf{Ang IMP}}_O)_{1\to2} = (\vec{H}_O)_2$$

$$\Sigma \bar{I}\vec{\omega}_1 + \Sigma m\vec{v}_1 \times \vec{r}_{G/O} + \int_{t_1}^{t_2} \vec{M}_O dt = \Sigma \bar{I}\vec{\omega}_2 + \Sigma m\vec{v}_2 \times \vec{r}_{G/O}$$

0.7 TEXTBOOKS AND REFERENCES

As this is a course companion, it is likely your instructor will assign another textbook. Below is a short list of especially well-written textbooks that are often adopted by instructors. While there are numerous original example problems within this course companion, many of the examples were inspired by problems written by others, especially from these four textbooks. Tables 0.3 and 0.4 are provided to both give appropriate attribution to the original inspiration and to direct instructors to similar problems for homework and exams. Students and instructors are encouraged to explore these problems for alternate arrangements, objectives, and solution approaches. Problems marked with an asterisk indicate that the example was inspired by it and/or is similar enough to be considered an adaptation. Below are the four book references.

[1] Beer, F. P., Johnston, E. R., Cornwell, P. J., and Self, B. P. 2018. *Vector Mechanics for Engineers: Dynamics*, 12th ed., McGraw-Hill Education, New York.

[2] Hibbeler, R. C. 2010. *Engineering Mechanics: Dynamics*, 12th ed., Prentice Hall, Upper Saddle River, NJ.

[3] Bedford, A. and Fowler, W. L. 2008. *Engineering Mechanics: Dynamics*, 5th ed., Pearson Prentice Hall, Upper Saddle River, NJ.

[4] Tongue, B. H. 2010. *Dynamics: Analysis and Design of Systems in Motion*, 2nd ed., John Wiley & Sons, Hoboken, NJ.

Table 0.3: Example problem reference correlation for Classes 13–18 (* indicates problem was inspired by)

Class 13					Class 14				
Ex.	[1]	[2]	[3]	[4]	Ex.	[1]	[2]	[3]	[4]
13.1	15.4	16-6	17.3		14.1	15.38	16-67	17.26	6.1.20
13.2	15.24	16-7	17.4	6.1.14	14.2	15.39	16-57	17.30	
13.3	15.31	16-13	17.6	6.1.47	14.3	15.62	16-70	17.31	6.1.44
13.4	15.9	16-19	17.5		14.4	15.64*	16-75	17.39	6.1.29
Class 15					Class 16				
Ex.	[1]	[2]	[3]	[4]	Ex.	[1]	[2]	[3]	[4]
15.1	15.74	16-90	17.80	6.2.10	16.1	15.117	16-114	17.85	6.3.9
15.2	15.83*	16-107	17.67	6.2.12	16.2	15.107	16-107	17.87	6.3.15
15.3	15.85	16-88	17.75	6.2.29	16.3	15.120	16-106	17.98	6.3.20
15.4	15.103	16-92	17.70	6.2.11	16.4	15.134	16-112	17.105	6.3.34
Class 15					Class 16				
Ex.	[1]	[2]	[3]	[4]	Ex.	[1]	[2]	[3]	[4]
17.1	15.135	16-93	17.106	6.3.32	18.1	15.CQ8	16-132	17.118	E6.16
17.2	15.145				18.2	15.153	16-139	17.123	6.4.28
17.3	SP15.16				18.3				
17.4					18.4	15.156		17.131	
Class 17					Class 18				
Ex.	[1]	[2]	[3]	[4]	Ex.	[1]	[2]	[3]	[4]
17.1	15.135	16-93	17.106	6.3.32	18.1	15.CQ8	16-132	17.118	E6.16
17.2	15.145				18.2	15.153	16-139	17.123	6.4.28
17.3	SP15.16				18.3				
17.4					18.4	15.156		17.131	

Table 0.4: Example problem reference correlation for Classes 19–24 and Appendix B (* indicates problem was inspired by)

Class 19					Class 20				
Ex.	[1]	[2]	[3]	[4]	Ex.	[1]	[2]	[3]	[4]
19.1	B.23	17–21	18.85	7.2.31	20.1	16.34*	17–79	18.16	7.2.55
19.2					20.2	16.61	17–117	18.9	7.3.17
19.3	B.3*	21–2	18.95	7.2.11	20.3	16.84	17–60	18.21	7.2.61
19.4	B.31	17–17	18.92	7.2.13	20.4	16.78	17–67		7.2.59
19.5	B.32	17–18	18.94	7.2.27					
Class 21					**Class 22**				
Ex.	[1]	[2]	[3]	[4]	Ex.	[1]	[2]	[3]	[4]
21.1	16.110	17–88	18.55	7.3.38	22.1	S17.1	18–8	19.36	7.5.3
21.2	SP16.12	17–93	18.66	7.3.26	22.2	17.25		19.20	
21.3	16.136		18.65		22.3	17.14	18–54	19.25	7.5.30
21.4	16.142		18.69		22.4	17.40	18–52	19.42	
					22.5	17.51		19.104	7.5.41
Class 23					**Class 24**				
Ex.	[1]	[2]	[3]	[4]	Ex.	[1]	[2]	[3]	[4]
23.1	17.60	19–7	19.96	7.4.12	24.1	17.97	19–30		7.4.32
23.2	SP17.9	19–33	19–59	7.4.19	24.2	17.127		19.72	7.4.37
23.3	17.87	19–35*	19.55	7.4.20	24.3	SP17.15	19–49	19.75	
23.4	17.78	19–23	19.57	7.4.21	24.4	17.129	19–43	19.74	
Appendix B.1					**Appendix B.2**				
Prob.	[1]	[2]	[3]	[4]	Prob.	[1]	[2]	[3]	[4]
B.1.1	15.31	16–10*	17.5	6.1.18	B.2.1	B.21	17–15	18.108	7.2.13
B.1.2	15.111	16–57	17.85	6.1.20	B.2.2	16.131	17–93	18.47	7.3.6
B.1.3	15.84	16–76	17.165	6.2.17	B.2.3	16.119	17–92*	18.46	7.3.7
B.1.4	15.94	16–87	17.78	6.2.23	B.2.4	17.14	18–54	19.29	7.5.30
B.1.5	15.98	16–92	17.77	6.2.9	B.2.5	17.28	18–40	19.20	7.5.37
B.1.6	15.123	16–103	17.93	6.3.15	B.2.6	17.130	19–43	19.75	7.4.37
B.1.7	15.125	16–106	17.97	6.3.26	B.2.7	17.138	19–31	19.72	–
B.1.8	15.124	16–110	17.95	6.3.30	B.2.8	17.121	19–49	19.64	–

OTHER REFERENCES:

Below are additional references used while writing Part 2.

- Newton, I. 1687. *Philosophiæ Naturalis Principia Mathematica.* DOI: 10.5479/sil.52126.39088015628399

- Meriam, J. L. and Kraige, L. G. 2012. *Engineering Mechanics: Dynamics*, vol. 2, John Wiley & Sons.

- Norton, R. L. 2020. *Design of Machinery*, 6th ed., McGraw-Hill.

- Myszka, David H. 2012. *Machines and Mechanisms: Applied Kinematic Analysis*, Pearson.

- Uicker, J. J., Pennock, G. R., and Shigley, J. E. 2011. *Theory of Machines and Mechanisms*, vol. 1, New York, Oxford University Press.

- To, Cho W. S. 2018. Engineering dynamics. *Synthesis Lectures on Mechanical Engineering 2.5*, pages 1–189. https://doi.org/10.2200/S00853ED1V01Y201805MEC015

- Nelson, E., Best, C. L., Best, C., McLean, W. G., McLean, W. G., and McLean, W. 1998. *Schaum's Outline of Engineering Mechanics*, McGraw Hill Professional.

- Farrow, W. C. and Weber, R. 1993. *Study Guide to Accompany Engineering Mechanics Dynamics*, Chichester, John Wiley.

- National Council of Examiners for Engineering, 2011. *Fundamentals of Engineering: Supplied-Reference Handbook*, Kaplan AEC Engineering.

- Diehl, E. J. 2018. Using cartoons to enhance engineering course concepts. *ASEE Annual Conference and Exposition.* https://peer.asee.org/authors/39810

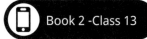

https://www.youtube.com/watch?v=FX0JQMPYtnI

Angular Kinematics of Rigid Body Motion

B.L.U.F. (Bottom Line Up Front)

- Rigid Body Motion (RBM) is a combination of translation and rotation.

- Rotation Kinematics has "Fundamental Angular Kinematic Relations."

  - Angular Position θ.
  - Angular Velocity $\omega = \frac{d\theta}{dt}$.
  - Angular Acceleration $\alpha = \frac{d\omega}{dt} = \frac{d^2\theta}{dt^2}$ and $\alpha = \omega \frac{d\omega}{d\theta}$.

- Constant Angular Velocity (speed) $\theta = \theta_0 + \omega t$.

- Constant Angular Acceleration $\omega = \omega_0 + \alpha t, \theta = \theta_0 + \omega_0 t + \frac{1}{2}\alpha t^2$, and $\omega^2 = \omega_0^2 + 2\alpha (\theta - \theta_0)$.

- These relationships are analogous to the translation relations introduced in Class 1 (vol. 1).

13.1   INTRODUCTION TO RIGID BODY MOTION KINEMATICS

Rigid Body Motion (RBM) differs from the motion of particles in that the object(s) analyzed have a shape that matters. Figure 13.1 shows Newtdog throwing a boomerang which we'll use to describe RBM. Note there are two points on the boomerang and the distance between these two points remains constant, therefore it is rigid. We use two points on each rigid body to keep track of the motion.

Figure 13.2a shows the boomerang in two dimensions with the two points labeled "A" and "B". The position vectors, $\vec{r}_A$ and $\vec{r}_B$, completely define the location of the object. We note that the relative position vector, $\vec{r}_{A/B}$, remains constant. This is the definition of "rigid" in the RBM, since if the relative position vector were to change the object would be flexible.

Figure 13.1: Newtdog throws a boomerang, a rigid body that moves with both translation and rotation (© E. Diehl).

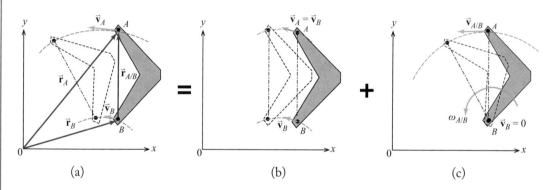

Figure 13.2: Rigid body motion (a) can be described as a combination of translation (b) and rotation (c).

Figure 13.2b shows translation of the boomerang while Figure 13.2c shows rotation of the body. The overall motion of the body is the combination of these two motions.

The conclusion we make is the most important thing to remember in RBM Kinematics:

"*Rigid Body Motion can be described as a combination of translation and rotation.*"

For particle kinematics we only needed translation to fully describe motion. We can use all of what we've learned about particle kinematics regarding translation and add rotation to the analysis. We begin by describing two special cases of RBM velocity: "pure translation" and "pure rotation."

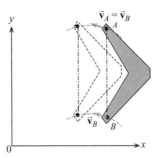

Figure 13.3: Pure translation of a rigid body, repeat of Figure 13.2b.

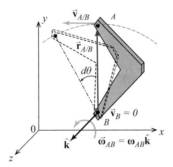

Figure 13.4: Pure rotation of a rigid body in one place.

13.2 PURE TRANSLATIONAL VELOCITY OR RIGID BODIES

When a rigid body translates but does not rotate we refer to this as "pure translation." On Figure 13.3 Points A and B are moving along parallel paths with equal magnitude velocity vectors. This is the definition of pure translation: every point on a rigid body has equal velocity (both magnitude and direction). We also note that the paths remain parallel, similar to an object riding on rails but without any turning.

13.3 PURE ROTATIONAL VELOCITY OF RIGID BODIES

When a rigid body rotates about a fixed axis we refer to it as having "pure rotation." Figure 13.4 shows the body rotating about Point B in the $x-y$ plane. The angular velocity of the body is thought of as being about an axis aligned with the $\hat{\mathbf{k}}$ unit vector. We use the lowercase Greek letter omega, ω, interchangeably with $\dot{\theta}$ to represent the angular speed. We note that the path of Point A is a circle centered at B. Also note that ω would be positive as drawn in Figure 13.4 since we use the right-hand rule where counter-clockwise is considered positive rotation.

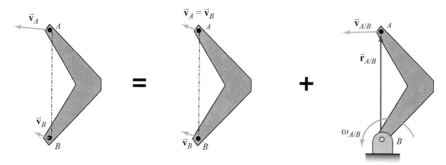

Figure 13.5: The velocity diagram for a velocity analysis.

The velocity of Point A, $\vec{\mathbf{v}}_A$, is also $\vec{\mathbf{v}}_{A/B}$ since B is fixed. We define this relative velocity using vector notation as:

$$\vec{\mathbf{v}}_{A/B} = \vec{\boldsymbol{\omega}}_{AB} \times \vec{\mathbf{r}}_{A/B}.$$

The cross-product produces a new vector which is perpendicular to the plane of the two crossed vectors. Since the angular velocity vector is coming out of the page, the velocity vector ends up perpendicular to the position vector and in the plane of the page. Note that A/B appears in the subscript of both the velocity and position vector. This can help make an association between these two.

13.4   VELOCITY DIAGRAM

To do an organized velocity analysis we will often draw a Velocity Diagram similar to Figure 13.5 which pictorially represents an object's motion broken down into translation and rotation. The drawing on the left is the complete motion. The second drawing describes the translation showing the velocity of Point B applied to Point A as well so it is pure translation. The third drawing uses Point B as the reference fixed point to describe the pure rotation portion of the motion.

In the next class/chapter we will use this diagram along with vector approaches to perform the velocity analysis in three separate ways. We will also use a similar approach beginning in Class 16 for acceleration analysis.

The remainder of this chapter and the examples which follow focus on the third drawing of the velocity diagram to use angular kinematics with rigid bodies in pure rotation.

13.5   FUNDAMENTAL ANGULAR KINEMATIC EQUATIONS

In Section 1.1 (vol. 1) we introduced the "Fundamental Kinematic Equations," but we only dealt with translation, so we may have wanted to refer to them as the "Fundamental Translational Kinematic Equations." These relate time, position, velocity, and acceleration to one another.

Table 13.1: Fundamental kinematic relations corollaries

	Angular Kinematics	Translational Kinematics
Position	θ	x, s, r, etc.
Velocity	$\omega = \dot{\theta} = \dfrac{d\theta}{dt}$	$v = \dot{x} = \dfrac{dx}{dt}$
Acceleration (function of time)	$\alpha = \ddot{\theta} = \dfrac{d\omega}{dt} = \dfrac{d^2\theta}{dt^2}$	$\alpha = \ddot{z} = \dfrac{dv}{dt} = \dfrac{d^2x}{dt^2}$
Acceleration (function of position)	$\alpha = \omega \dfrac{d\omega}{d\theta}$	$\alpha = v \dfrac{dv}{dx}$

We now introduce the "Fundamental Angular Kinematic Equations" to relate time, *angular* position, *angular* velocity, and *angular* acceleration. It's convenient to think of these as analogous to their translational counterparts when applying them to solve problems. All of the problem types presented in Classes 1 (vol. 1) and 2 (vol. 1) (integration, differentiation, and the special cases) are applicable when solving problems with angular kinematics.

We'll stick with two-dimensional motion and discuss the scalar versions of the equations in order to keep the presentation somewhat simplified.

Position is defined by the angle theta, θ, measured counter-clockwise from the positive x-axis. The angular velocity is the time rate change of angular position: $\omega = \frac{d\theta}{dt}$ also written as $\dot{\theta}$. Angular acceleration (designated with Greek lowercase alpha, α) is the time rate change of angular velocity and written as $\alpha = \frac{d\omega}{dt} = \frac{d^2\theta}{dt^2}$. Theta double-dot ($\ddot{\theta}$) is also used interchangeably with alpha (α). As with translational acceleration there is an alternative angular acceleration which is useful when its function is related to position rather than time: $\alpha = \omega \frac{d\omega}{d\theta}$. We know this to be true because we can rewrite $\omega = \frac{d\theta}{dt}$ as $dt = \frac{d\theta}{\omega}$ and substitute it into $\alpha = \frac{d\omega}{dt}$. The corollary relationships between angular and translational kinematics are summarized in Table 13.1.

13.6   ANGULAR KINEMATICS SPECIAL CASES

Just as with translational kinematics we often have special cases where the angular velocity is constant or the angular acceleration is constant. Table 13.2 presents the corollary equations.

13.7   IMPORTANT IDEAS ABOUT VELOCITY OF RIGID BODIES

Quite often when doing a two-dimensional analysis of a pure rotational rigid body, it is convenient to simply find the magnitude of $\vec{v}_{A/B} = \vec{\omega}_{AB} \times \vec{r}_{A/B}$ by writing $v_{A/B} = \omega_{AB}\, r_{A/B}$ as represented in Figure 13.6. We mention this here because we'll use this simple relationship often

Table 13.2: Uniform motion kinematic special cases corollaries

	Angular Kinematics	**Translational Kinematics**
Position from Constant Velocity	$\theta = \theta_0 + \omega_c t$	$x = x_0 + v_c t$
Velocity from Constant Acceleration	$\omega = \omega_0 + \alpha_c t$	$v = v_0 + a_c t$
Position from Constant Acceleration	$\theta = \theta_0 + \omega_0 t + \frac{1}{2} \alpha_c t^2$	$x = x_0 + v_0 t + \frac{1}{2} a_c t^2$
Velocity from Constant Acceleration	$\omega^2 = \omega_0^2 + 2\alpha_c(\theta - \theta_0)$	$v^2 = v_0^2 + 2a_c(x - x_0)$

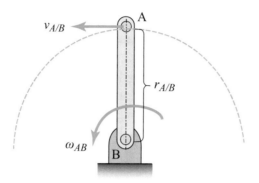

Figure 13.6: Velocity due to pure rotational RBM.

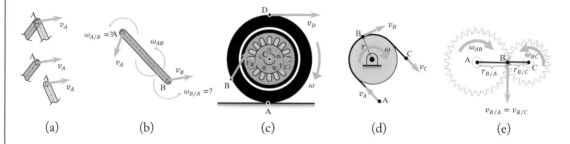

Figure 13.7: Some velocity analysis concepts.

and some students seem to overlook it when it's presented in the vector form. Note the path in this pinned link is a circle and, as we know, the velocity is tangent to that path.

Figure 13.7 presents some velocity analysis concepts we'll use in examples in this and later classes.

Attached Points   Attached points of two rigid bodies (such as with a pinned joint in Figure 13.7a) have the same velocity (and the same acceleration). This might seem obvious, but we'll see as we do RBM problems involving multiple parts that this logic allows us to numerically link components.

Is Angular Velocity about a Specific Point?   While we often will say an object rotates about a particular point, the angular velocity is the same everywhere on the body and isn't specific to that point as indicated in Figure 13.7b.

No Slip Wheels   The tire shown in Figure 13.7c has a point in contact with the ground. If there is no slipping of the tire (no pealing out or skidding) the velocity at the point of the tire that touches the ground equals the velocity of the ground: zero. In Class 15 we will see that this point is called the "Instantaneous Center of Rotation" and is useful in finding the velocity at any point on the tire because angular velocity isn't specifically about any point.

No Slip Belts and Pulleys   In Figure 13.6d we have a belt on a pulley that doesn't slip. Since there is no slip, the velocities at any point where the belt touches the pulley are the same. Once the belt leaves the pulley it has the same velocity as on the outside of the pulley since we assume the belt doesn't stretch. Again this may seem obvious, but knowing this will help us walk through the logic of solving some problems.

Gears   Gears have teeth that come in contact periodically. There is an imaginary circle on each gear called a "pitch circle" and the point at which pitch circles of two gears touch is the "pitch point." The velocity at this point is the same for both gears because they too cannot slip (without breaking off teeth). As shown in Figure 13.7e, we can find the rotational velocity relationship based on this common velocity, so $v_{B/A} = \omega_{AB} \, r_{B/A} = v_{B/C} = \omega_{BC} \, r_{B/C}$. We can see that the angular velocity is inversely proportional to the radius (and diameter) of the gears: $\omega_{BC} = \omega_{AB} \frac{r_{B/A}}{r_{B/C}}$. This is a useful concept for gear train applications.

13.8   ACCELERATION OF RIGID BODIES IN PURE ROTATION

The acceleration of points on rigid bodies in pure rotation are readily described using concepts we covered in Classes 4 (vol. 1) and 5 (vol. 1) on Particle Kinematics. We use the terminology of path coordinates which has normal and tangential components. In Figure 13.8 we show a generic link in pure rotation. The acceleration of Point A, which follows a circular path, is made up of two parts: normal acceleration due to the change in direction and tangential acceleration due to an increase in speed.

The normal acceleration component is:

$$\boxed{(a_A)_n = \omega_{AB}{}^2 r_{A/B}} \, .$$

We note that this equation is related to the normal acceleration component we used in particle kinematics, $a_n = \frac{v^2}{\rho}$, because $\rho = r_{A/B}$ for this link, and $v = \omega_{AB} r_{A/B}$. If we substitute these,

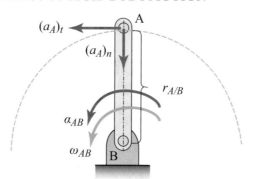

Figure 13.8: Acceleration of rigid body in pure rotation.

$a_n = \frac{(\omega_{AB} r_{A/B})^2}{r_{A/B}}$, we arrive at the same result. It is also similar to part of the polar coordinates radial components: $r\dot{\theta}^2$. We recall that the full radial component in polar coordinates was $a_r = \ddot{r} - r\dot{\theta}^2$, but since the length of the link doesn't change (it's rigid) $\ddot{r} = 0$.

The tangential acceleration component is:

$$\boxed{(a_A)_t = \alpha_{AB} r_{A/B}} \ .$$

In this case the tangential component is similar to a part of the transverse component in polar coordinates: $r\ddot{\theta}$. We recall that the full transverse component in polar coordinates is $a_\theta = r\ddot{\theta} + 2\dot{r}\dot{\theta}$, but again since the length doesn't change $\dot{r} = 0$.

When looking for the total acceleration of the end point in a Rigid Body in Pure Rotation, we write:

$$a_A = \sqrt{(a_A)_t^2 + (a_A)_n^2} = \sqrt{\left(\alpha_{AB} r_{A/B}\right)^2 + \left(\omega_{AB}^2 r_{A/B}\right)^2}.$$

We will formalize the normal and tangential equations in vector form in Class 16 when we begin Rigid Body Acceleration consisting of both translation and rotation.

Example 13.1

A small engine (Figure 13.9) is running at its idle speed of 500 rpm counter-clockwise when the throttle is increased to full speed of 3600 rpm which takes 10 s. The engine runs for 30 s at full speed before it is turned off and coasts to a rest in 45 s. Determine the total number of revolutions the engine rotates from throttle increase to stopping.

Idle speed:

$$\omega = \frac{(500 \text{ rpm}) (2\pi \text{ rad/rev})}{(60 \text{ s/min})} = 52.36 \text{ rad/s}.$$

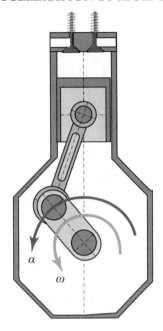

Figure 13.9: Example 13.1.

Full speed:

$$\omega = \frac{(3600 \text{ rpm}) (2\pi \text{ rad/rev})}{(60 \text{ s/min})} = 377.0 \text{ rad/s}.$$

Acceleration idle to full speed:

$$\omega = \omega_0 + \alpha t \quad \alpha = \frac{\omega - \omega_0}{t} = \frac{(377.0) - (52.36)}{(10)} = 32.46 \text{ rad/s}^2.$$

Angular position when changing from idle to full speed:

$$\theta_1 = \theta_0 + \omega_0 t + \frac{1}{2}\alpha t^2 = (0) + (52.36)(10) + \frac{1}{2}(32.46)(10)^2 = 2{,}147 \text{ rad}.$$

Angular position after running at constant full speed:

$$\theta_2 = \theta_1 + \omega t = (2{,}147) + (377.0)(30) = 13{,}460 \text{ rad}.$$

Deceleration coasting to stop:

$$\omega = \omega_0 + \alpha t \quad \alpha = \frac{\omega - \omega_0}{t} = \frac{(0) - (377.0)}{(45)} = -8.378 \text{ rad/s}^2.$$

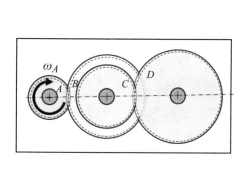
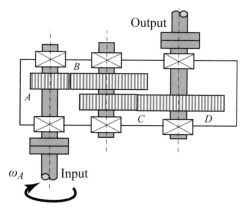

Figure 13.10: Example 13.2.

Angular position after coasting to stop:

$$\theta_3 = \theta_2 + \omega_0 t + \frac{1}{2}\alpha t^2 = (13{,}460) + (377.0)(45) + \frac{1}{2}(-8.378)(45)^2 = 21{,}940 \text{ rad.}$$

Total Revolutions:

$$\theta = \frac{(21{,}940 \text{ rad})}{(2\pi \text{ rad/rev})} = 3{,}492 \text{ rev}$$

$$\boxed{\theta = 3{,}490 \text{ rev}}.$$

We could have also used $\omega^2 = \omega_0^2 + 2\alpha_c(\theta - \theta_0)$ to find the number of radians used between speeds, for instance in the acceleration from idle to full speed:

$$(377.0)^2 = (52.36)^2 + 2(32.46)(\theta_1 - (0)),$$

$\theta_1 = 2{,}147$ rad which is the same result we had previously.

Example 13.2
The two-stage gear train shown in Figure 13.10 is used to reduce the output shaft speed for an engine operating at 1500 rpm. The gear diameters are $d_A = 80$ mm, $d_B = 160$ mm, $d_C = 120$ mm, and $d_D = 180$ mm. Gears B and C attached to the intermediate shaft must turn together. Determine the speeds of the intermediate and output shafts.

Angular speed of A:

$$\omega_A = \frac{(1500 \text{ rpm})(2\pi \text{ rad/rev})}{(60 \text{ s/min})} = 157.1 \text{ rad/s.}$$

We can break the analysis down by the gear pairs as in Figure 13.11.

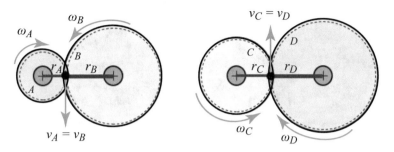

Figure 13.11: Velocities of gears in Example 13.2.

Velocity at point shared by A and B:

$$v_A = v_B = r_A \omega_A = \frac{(0.080)}{2} (157.1) = 6.283 \text{ m/s}.$$

Angular speed of B:

$$v_B = r_B \omega_B \quad \omega_B = \frac{v_B}{r_B} = \frac{(6.283)}{(0.160)/2} = 78.54 \text{ rad/s}.$$

$$\omega_B = \frac{(78.54 \text{ rad/s}) (60 \text{ s/min})}{(2\pi \text{ rad/rev})} = 750.0 \text{ rpm} \circlearrowleft .$$

Since B and C are on the same shaft and rotate together:

$$\omega_C = \omega_B = 78.54 \text{ rad/s}.$$

Velocity at point shared by C and D:

$$v_C = v_D = r_C \omega_C = \frac{(0.120)}{2} (78.54) = 4.712 \text{ m/s}.$$

Angular speed of D:

$$v_D = r_D \omega_D \quad \omega_D = \frac{v_D}{r_D} = \frac{(4.712)}{(0.180)/2} = 52.36 \text{ rad/s}$$

$$\omega_D = \frac{(52.36 \text{ rad/s}) (60 \text{ s/min})}{(2\pi \text{ rad/rev})} = 500.0 \text{ rpm} \circlearrowright$$

$$\boxed{\omega_B = 750 \text{ rpm} \circlearrowleft \text{ rpm}} \quad \boxed{\omega_D = 500 \text{ rpm} \circlearrowright} .$$

We can see that each stage reverses the direction of rotation. The overall gear ratio is 1:3 reducing. This calculation could have been done in one step, omitting unit conversions and without changing diameter to radius, as these cancel out. So:

$$\omega_D = \frac{d_A}{d_B} \frac{d_C}{d_D} \omega_A = \frac{(80)}{(160)} \frac{(120)}{(180)} (1500) = 500.0 \text{ rpm}.$$

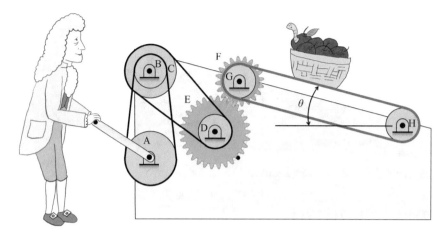

Figure 13.12: Example 13.3 (© E. Diehl).

Example 13.3

Newtdog cranks an unnecessarily complicated conveyor belt machine (Figure 13.12) to move Wormy's basket of apples ($W = 30$ lb) upward at an angle of $\theta = 20°$. He starts turning the crank of the stopped machine, increasing his cranking speed for one minute until the basket is going up the belt at 3 ft/s without slipping. The pulley dimensions are $r_A = r_C = 9$ in and $r_B = r_D = r_G = r_H = 4$ in, and the gear dimensions are $d_E = 20$ in and $d_F = 12$ in. The friction of the system causes an efficiency of $\eta = 25\%$. Determine Newtdog's full speed cranking and direction when he finishes, the total acceleration magnitude of a point on gear E just before he reaches full speed, and the power Newtdog is applying at the crank.

This type of problem requires us to use multiple concepts. So let's work backward and see where we go.

The final velocity of Wormy's basket is $v = 3$ ft/s which takes $t = 60$ s. The acceleration of the conveyor belt is:

$$v = v_0 + at.$$

$$a = \frac{v - v_0}{t} = \frac{(3) - (0)}{(60)} = 0.05 \text{ ft/s}^2.$$

We can start walking through the mechanism backward. The conveyor belt speed and acceleration are applied to a sketch of the elements separated (Figure 13.13). The included arrows assist in visualizing the direction of each component.

The conveyor belt speed and acceleration are the same as a point on pulley G so:

$$v_G = 3 \text{ ft/s} \qquad a_G = 0.05 \text{ ft/s}^2.$$

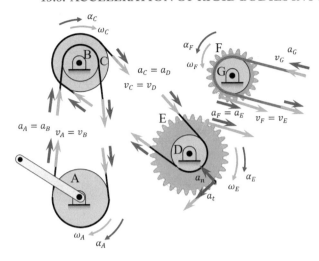

Figure 13.13: Component velocities and accelerations of Example 13.3.

Pulley G and gear F angular speed and acc:

$$\omega_G = \omega_F = \frac{v_G}{r_G} = \frac{(3)}{(9)/12} = 4.000 \text{ rad/s}$$

$$\alpha_G = \alpha_F = \frac{a_G}{r_G} = \frac{(0.05)}{(9)/12} = 0.06667 \text{ rad/s}^2.$$

Gear F and gear E teeth speed and acc:

$$v_F = v_E = \omega_F r_F = (4.000)\frac{(12)}{(12)(2)} = 2.000 \text{ ft/s}$$

$$a_F = a_E = \alpha_F r_F = (0.06667)\frac{(12)}{(12)(2)} = 0.03333 \text{ ft/s}^2.$$

Gear E and pulley D angular speed and acc:

$$\omega_E = \omega_D = \frac{v_E}{r_E} = \frac{(2.000)}{\dfrac{(20)}{(12)(2)}} = 2.400 \text{ rad/s}$$

$$\alpha_E = \alpha_D = \frac{a_E}{r_E} = \frac{(0.03333)}{\dfrac{(20)}{(12)(2)}} = 0.04000 \text{ rad/s}^2.$$

Pulleys D and C belt speed and acc:

$$v_D = v_C = \omega_D r_D = (2.400)(4)/12 = 0.8000 \text{ ft/s}$$

$$a_D = a_C = \alpha_D r_D = (0.04000)\,(4)/12 = 0.01333 \text{ ft/s}^2.$$

Pulleys C and B angular speed and acc:

$$\omega_C = \omega_B = \frac{v_C}{r_C} = \frac{(0.8000\ )}{(9)/12} = 1.067 \text{ rad/s}$$

$$\alpha_C = \alpha_B = \frac{a_C}{r_C} = \frac{(0.01333)}{(9)/12} = 0.01778 \text{ rad/s}^2.$$

Pulleys A and B belt speed and acc:

$$v_B = v_A = \omega_B r_B = (1.067)\,(4)/12 = 0.3556 \text{ ft/s}$$

$$a_B = a_A = \alpha_B r_B = (0.01778)\,(4)/12 = 0.005926 \text{ ft/s}^2.$$

Pulley A angular speed and acc:

$$\omega_A = \frac{v_A}{r_A} = \frac{(0.3556)}{(9)/12} = 0.4741 \text{ rad/s} = \frac{(0.4741 \text{ rad/s})\,(60 \text{ s/m})}{(2\pi \text{ rad/s})} = 4.527 \text{ rpm}$$

$$\alpha_A = \frac{a_A}{r_A} = \frac{(0.005926)}{(9)/12} = 0.007901 \text{ rad/s}^2.$$

Newtdog must crank the handle:

$$\boxed{\omega_A = 4.53 \text{ rpm } \circlearrowleft} \quad and \quad \boxed{\alpha_A = 0.00790 \text{ rad/s}^2 \ \circlearrowleft}.$$

The total acceleration of a point on Gear E:

$$(a_E)_n = \omega_E{}^2 r_E = (2.400)^2 \frac{(20)}{(12)\,(2)} = 4.800 \text{ ft/s}^2$$

$$(a_E)_t = \alpha_E r_E = (0.04000)\,\frac{(20)}{(12)\,(2)} = 0.03333 \text{ ft/s}^2 \quad \text{(already found as ``}a_E\text{'')}$$

$$a_{E,total} = \sqrt{(a_E)_t^2 + (a_E)_n^2} = \sqrt{(0.03333)^2 + (4.800)^2} = 4.800 \text{ ft/s}^2$$

$$\boxed{a_{E,total} = 4.80 \text{ ft/s}^2}.$$

We note that the normal acceleration is so much larger than the tangential that there's no practical difference between it and total acceleration.

To get the power we will find the force applied by the conveyor belt onto the basket. This is a frictional force that must address that the basket is accelerating, so we need an FBD/IBD pair.

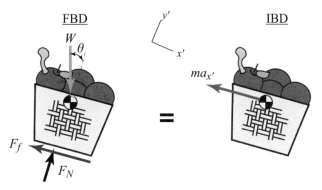

Figure 13.14: FBD/IBD pair of Example 13.3 (© E. Diehl).

$$\searrow \sum F_{x'} = ma_{x'}$$

$$- F_f + W \sin \theta = -ma_{x'}$$

$$F_f = ma_{x'} + W \sin \theta = \left(\frac{30}{32.2}\right)(0.05) + (30)\sin\left(20°\right) = 10.31 \text{ lb}.$$

The power required at the basket is:

$$\mathbb{P} = F \, v = \frac{(10.31 \text{ lb}) (3 \text{ ft/s})}{\left(550 \, \dfrac{\text{ft} \cdot \text{lb/s}}{\text{hp}}\right)} = 0.05622 \text{ hp}.$$

The powered required at the crank:

$$\eta_{ovr} = \frac{WYW}{WYPF} = \frac{\mathbb{P}_{WYW}}{\mathbb{P}_{WYPF}} = \frac{(0.05622)}{\mathbb{P}_{WYPF}} = 0.25 \quad \mathbb{P}_{WYPF} = 0.2249 \text{ hp}$$

$$\boxed{\mathbb{P} = 0.225 \text{ hp}}.$$

Example 13.4

The angular acceleration of a turbine due to aerodynamic drag on the rotating surfaces is empirically determined to equal $\alpha = \left(1.25 \cdot 10^{-3}\right) \omega^2$. How much time does it take for the turbine to change speed from 100 rpm to 500 rpm, and how may revolutions?

The turbine speeds converted into radians per second:

$$\omega_1 = \frac{(100 \text{ rpm}) (2\pi \text{ rad/rev})}{(60 \text{ s/min})} = 10.47 \text{ rad/s}$$

$$\omega_2 = \frac{(500 \text{ rpm}) (2\pi \text{ rad/rev})}{(60 \text{ s/min})} = 52.36 \text{ rad/s}.$$

We recognize that the angular acceleration is a function of angular speed, and since we are first interested in time we use the following:

$$\alpha = \frac{d\omega}{dt}$$

$$dt = \frac{1}{\alpha} d\omega$$

$$\int_0^t dt = \int_{\omega_1}^{\omega_2} \frac{1}{\alpha} d\omega$$

$$\int_0^t dt = t = \int_{\omega_1}^{\omega_2} \frac{1}{\left(1.25 \cdot 10^{-3}\right) \omega^2} d\omega = -\frac{1}{\left(1.25 \cdot 10^{-3}\right) \omega}\bigg|_{\omega_1}^{\omega_2} = -\frac{1}{\left(1.25 \cdot 10^{-3}\right)} \left(\frac{1}{\omega_2} - \frac{1}{\omega_1}\right)$$

$$t = -\frac{1}{\left(1.25 \cdot 10^{-3}\right)} \left(\frac{1}{(52.36)} - \frac{1}{(10.47)}\right) = 61.13 \text{ s}$$

$$\boxed{t = 61.1 \text{ s}}.$$

To find the number of revolutions we can use the other acceleration relation:

$$\alpha = \omega \frac{d\omega}{d\theta}$$

$$d\theta = \frac{\omega}{\alpha} d\omega$$

$$\int_0^\theta d\theta = \int_{\omega_1}^{\omega_2} \frac{\omega}{\alpha} d\omega$$

$$\int_0^\theta d\theta = \theta = \int_{\omega_1}^{\omega_2} \frac{\omega}{\left(1.25 \cdot 10^{-3}\right) \omega^2} d\omega = \int_{\omega_1}^{\omega_2} \frac{1}{\left(1.25 \cdot 10^{-3}\right) \omega} d\omega = \frac{1}{\left(1.25 \cdot 10^{-3}\right)} \ln \omega \bigg|_{\omega_1}^{\omega_2}$$

$$\theta = \frac{1}{\left(1.25 \cdot 10^{-3}\right)} (\ln \omega_2 - \ln \omega_1) = \frac{1}{\left(1.25 \cdot 10^{-3}\right)} (\ln (52.36) - \ln (10.47)) = 1{,}288 \text{ rad}$$

$$\theta = \frac{(1{,}288 \text{ rad})}{(2\pi \text{ rad/rev})} = 204.9 \text{ rev}$$

$$\boxed{\theta = 205 \text{ rev}}\;.$$

This last example is included to re-emphasize that acceleration isn't always constant, even with angular acceleration. Note that this is a somewhat realistic scenario. Since drag force is proportional to the fluid velocity squared and remembering force is also proportional to acceleration, a proportional relation between angular acceleration and angular speed squared might also be anticipated.

Book 2 - Class 14

https://www.youtube.com/watch?v=FefA1XhOVj8

CLASS 14

Absolute and Relative Velocity

B.L.U.F. (Bottom Line Up Front)

- The Velocity Diagram of Rigid Body Motion (RBM) depicts a combination of translation and rotation.

- The Velocity Diagram is also a graphical representation of relative motion between two points on a rigid body.

- Either point on an object represented in a Velocity Diagram can be used as the reference.

$$\vec{v}_A = \vec{v}_B + \vec{v}_{A/B}$$
$$\vec{v}_B = \vec{v}_A + \vec{v}_{B/A}$$

- There are several approaches to solving a velocity analysis using the Velocity Diagram.

14.1 RELATIVE VELOCITY ON A RIGID BODY

In Class 3 (vol. 1) we discussed relative motion using vectors. This concept is useful to describe rigid body motion. The two points on a rigid body move together but also relative to one another (in rotation only). We select one point as the translational reference and identify how the other point rotates about it (its relative motion). We write the relative motion equation for velocity as:

$$\vec{v}_A = \vec{v}_B + \vec{v}_{A/B}.$$

Written here $\vec{v}_B$ for Point B is the reference and $\vec{v}_{A/B}$ is the relative motion of A with respect to B. Together they are used to describe the velocity of A, $\vec{v}_A$. We will also use this concept in acceleration analysis.

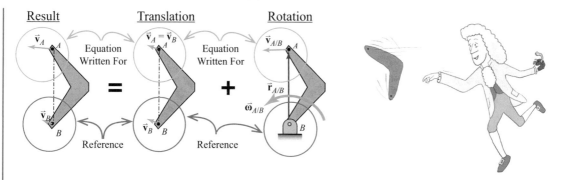

Figure 14.1: Velocity diagram breaks RBM into translation and rotation using one point as a reference (© E. Diehl).

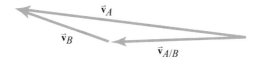

Figure 14.2: Vector triangle of velocity relative motion equation.

14.2   APPLYING THE RIGID BODY MOTION KINEMATICS VELOCITY DIAGRAM

In Class 13 we introduced the Velocity Diagram that we will use for analysis of Rigid Body Motion (RBM). Figure 14.1 is a repeat of 13.2 but with emphasis on how this diagram relates to the velocity relative motion equation.

The velocity relative motion equation has three parts, just as the Velocity Diagram has three parts: Result = Translation + Rotation. The equation is written for Point A and describes it. Point B is the reference for the pure translation portion of the motion. We describe the pure rotation of the boomerang as if it were pinned at the reference Point B:

$$\underbrace{\vec{v}_A}_{\text{Result}} = \underbrace{\vec{v}_B}_{\substack{\text{Translation} \\ \text{(Reference)}}} + \underbrace{\vec{v}_{A/B}}_{\substack{\text{Rotation} \\ \text{(Relative Motion)}}} .$$

Figure 14.2 shows the vector triangle formed from this breakdown of motion. The vector magnitudes are enlarged here to help visualize the triangle formed in the relative motion equation.

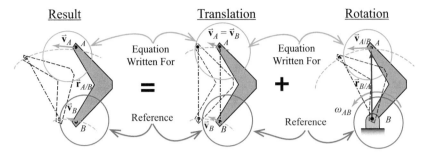

Figure 14.3: Velocity diagram using Point B as the reference.

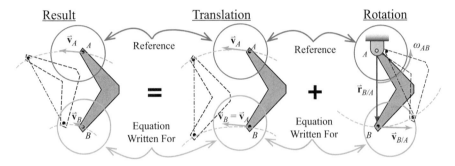

Figure 14.4: Velocity diagram using Point A as the reference.

14.2.1   CHOOSING A REFERENCE POINT

Either point on a rigid body can be used as the reference and the other point is the "result." Figure 14.3 shows the Velocity Diagram using Point B as the reference ($\vec{v}_A = \vec{v}_B + \vec{v}_{A/B}$) and Figure 14.4 shows Point A as the reference ($\vec{v}_B = \vec{v}_A + \vec{v}_{B/A}$). There are often solution strategy advantages to using one point rather than the other, especially when we introduce the acceleration version of this approach. In the sections which follow we'll demonstrate whether either reference is possible and use several different solution approaches: Vector Math, Vector Triangles, and Vector Components.

14.2.2   SOLUTION METHOD USING VECTOR MATH

We can replace the relative motion portion of the velocity equation with the cross product of angular velocity and relative position for pure rotation we introduced in Class 13: $\vec{v}_{A/B} = \vec{\omega}_{AB} \times \vec{r}_{A/B}$:

$$\vec{v}_A = \vec{v}_B + \vec{\omega}_{AB} \times \vec{r}_{A/B}.$$

We are restricting our discussion to two-dimensions and using Cartesean Coordinates, but the above equation remains valid for three dimensions and any coordinate system. For two dimen-

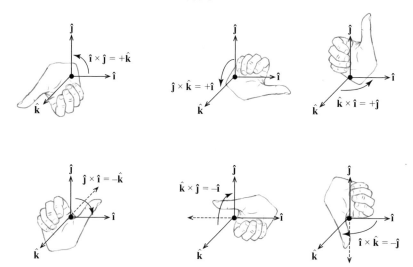

Figure 14.5: Cross product results based on right-hand rule.

sions each vector in the above is written as:

$$\vec{v}_A = (v_A)_x\,\hat{\imath} + (v_A)_y\,\hat{\jmath}$$

$$\vec{v}_B = (v_B)_x\,\hat{\imath} + (v_B)_y\,\hat{\jmath}$$

$$\vec{\omega}_{AB} = \omega_{AB}\hat{k}$$

$$\vec{r}_{A/B} = (r_{A/B})_x\,\hat{\imath} + (r_{A/B})_y\,\hat{\jmath}.$$

One method of determining/remembering the results of a cross-product of two vectors is to follows the right-hand rule convention. Figure 14.5 illustrates the six possible combinations of cross-product applied to unit vectors. The fingers of the right-hand curl from the first unit vector toward the second unit vector in the cross-product, and the thumb points toward the resulting new unit vector including direction.

For instance, the cross product of $\hat{\imath}$ to $\hat{\jmath}$ will produce $+\hat{k}$ since the fingers curl counterclockwise and the thumb points in the new vector direction as in the upper left of Figure 14.5. If we instead take the cross product of $\hat{\jmath}$ to $\hat{\imath}$ we produce $-\hat{k}$ since the fingers curl clockwise and the thumb points in that direction. This is consistent with our convention of treating counterclockwise as positive rotation.

Re-writing the velocity equation for two dimensions we get:

$$(v_A)_x\,\hat{\imath} + (v_A)_y\,\hat{\jmath} = (v_B)_x\,\hat{\imath} + (v_B)_y\,\hat{\jmath} + \omega_{AB}\hat{k} \times \left((r_{A/B})_x\,\hat{\imath} + (r_{A/B})_y\,\hat{\jmath}\right)$$

$$(v_A)_x\,\hat{\imath} + (v_A)_y\,\hat{\jmath} = (v_B)_x\,\hat{\imath} + (v_B)_y\,\hat{\jmath} + \omega_{AB}\hat{k} \times (r_{A/B})_x\,\hat{\imath} + \omega_{AB}\hat{k} \times (r_{A/B})_y\,\hat{\jmath}$$

$$(v_A)_x\,\hat{\imath} + (v_A)_y\,\hat{\jmath} = (v_B)_x\,\hat{\imath} + (v_B)_y\,\hat{\jmath} + \omega_{AB}\,(r_{A/B})_x\,\underbrace{\hat{k}\times\hat{\imath}}_{+\hat{\jmath}} + \omega_{AB}\,(r_{A/B})_y\,\underbrace{\hat{k}\times\hat{\jmath}}_{-\hat{\imath}}$$

$$\vec{\mathbf{v}}_A = \left[(v_B)_x - \omega_{AB}\,(r_{A/B})_y\right]\hat{\imath} + \left[(v_B)_y + \omega_{AB}\,(r_{A/B})_x\right]\hat{\jmath}.$$

We can write this as components

$$(v_A)_x = (v_B)_x - \omega_{AB}(r_{A/B})_y \quad \text{and} \quad (v_A)_y = (v_B)_y + \omega_{AB}(r_{A/B})_x.$$

Note that the y-component of the radius is used to find the x-component of the relative velocity and vice versa. Rather than memorize this result, it is best to remember the concepts and apply them to each new scenario. We'll see similar results using two other methods. The Vector Diagram isn't necessary for this approach but helps to confirm the resulting signs/directions of the cross product.

For completeness, we can demonstrate that using Point A as the reference instead of Point B yields the same result:

$$\vec{\mathbf{v}}_B = \vec{\mathbf{v}}_A + \vec{\mathbf{v}}_{B/A} = \vec{\mathbf{v}}_A + \vec{\omega}_{AB} \times \vec{\mathbf{r}}_{B/A}.$$

Written this way the position vector direction is reversed:

$$\vec{\mathbf{r}}_{B/A} = -\vec{\mathbf{r}}_{A/B} = -(r_{A/B})_x\,\hat{\imath} - (r_{A/B})_y\,\hat{\jmath}$$

$$(v_B)_x\,\hat{\imath} + (v_B)_y\,\hat{\jmath} = (v_A)_x\,\hat{\imath} + (v_A)_y\,\hat{\jmath} + \omega_{AB}\hat{k} \times \left(-(r_{A/B})_x\,\hat{\imath} - (r_{A/B})_y\,\hat{\jmath}\right)$$

$$(v_B)_x\,\hat{\imath} + (v_B)_y\,\hat{\jmath} = (v_A)_x\,\hat{\imath} + (v_A)_y\,\hat{\jmath} - \omega_{AB}\,(r_{A/B})_x\,\underbrace{\hat{k}\times\hat{\imath}}_{+\hat{\jmath}} - \omega_{AB}\,(r_{A/B})_y\,\underbrace{\hat{k}\times\hat{\jmath}}_{-\hat{\imath}}$$

$$(v_B)_x = (v_A)_x + \omega_{AB}(r_{A/B})_y \quad \text{and} \quad (v_B)_y = (v_A)_y - \omega_{AB}(r_{A/B})_x.$$

This is the same result as we found using B as the reference/translational term.

14.2.3   SOLUTION METHOD USING VECTOR TRIANGLES

We recall from the vector addition review in Class 3 (vol. 1) that vector triangles are formed when the added vectors are arranged head-to-tail and the results found from tail-tail and head-head. Figure 14.6a repeats the vector triangle from Figure 14.2 for comparison to Figure 14.6b which uses Point A as the reference instead of Point B. This again demonstrates the validity of using either point as the reference for translation.

To demonstrate application of the vector triangle approach we present a generic velocity vector triangle in Figure 14.7 that represents $\vec{\mathbf{v}}_A = \vec{\mathbf{v}}_B + \vec{\mathbf{v}}_{A/B}$. In a typical scenario we might know or be able to find the magnitudes and directions of $\vec{\mathbf{v}}_B$ and $\vec{\mathbf{v}}_{A/B}$. The magnitude of $\vec{\mathbf{v}}_B$ is

Figure 14.6: Velocity vector triangles for (a) Point B as the reference and (b) Point A as the reference.

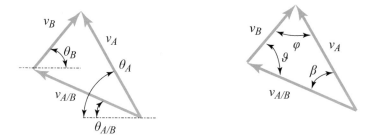

Figure 14.7: Generic velocity vector triangles.

v_B and the angle is $\theta_B = \tan \frac{(v_B)_y}{(v_B)_x}$. The magnitude of $\vec{v}_{A/B}$ is found from $v_{A/B} = \omega_{AB}\, r_{A/B}$ and the direction (we'll designate this as $\theta_{A/B}$ even though it is not a relative angle) is perpendicular to the position vector.

We use triangle geometry to implement the vector triangle method. To do this we need to know the interior angles, so we designate them with arbitrary Greek letters, β, ϑ, and φ. From simple geometry we know $\vartheta = \theta_B + \theta_{A/B}$. We can't find the other angles without knowing θ_A, but this is part of the solution, so we'll have to make do. In this scenario we know three triangle dimensions: v_B, $v_{A/B}$, and ϑ. With these we can find the unknown side (v_A) and angles (β and φ) using the Law of Sines (LoS) and Law of Cosines (LoC).

Since we have two sides and the angle in between, we can apply LoC:

$$v_A = \sqrt{v_B^2 + v_{A/B}^2 - 2v_B v_{A/B} \cos \vartheta}.$$

With this magnitude we can find the other two angles using LoS:

$$\frac{v_A}{\sin \vartheta} = \frac{v_B}{\sin \beta}.$$

The angle of the desired result is $\theta_A = \theta_{A/B} + \beta$. With this we know the magnitude and direction of the desired velocity vector: $\vec{v}_A = v_A \nearrow \theta_A$.

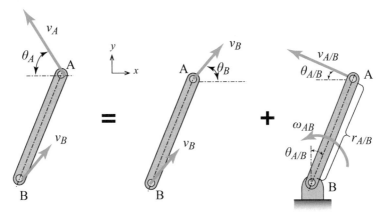

Figure 14.8: Generic velocity diagram of a link.

Often the challenging aspect to this approach is keeping track of the angles. This can be greatly assisted by drawing a careful sketch using a straight edge on graph paper approximately to scale. Often it is good practice to use the vector triangle approach to confirm or check the results of another method. Velocity triangles work well for velocity analysis but can be very confusing when applied to acceleration analysis, as discussed in Section 16.4.

14.2.4   SOLUTION METHOD USING VECTOR COMPONENTS OF VELOCITY DIAGRAM

In Section 13.4 we introduced the Velocity Diagram which breaks rigid body motion into translation and rotation. Figure 14.8 illustrates a generic link problem that coincides with the vectors used in the previous demonstration of the vector triangle approach. The Vector Diagram is especially useful for breaking down the vector components to find the unknown values.

We write out the equation and then break it into components assigning signs based on directions indicated in the diagrams:

$$\vec{\mathbf{v}}_A = \vec{\mathbf{v}}_B + \vec{\mathbf{v}}_{A/B}.$$

x-dir:

$$(v_A)_x = (v_B)_x - (v_{A/B})_x$$

$$(v_A)_x = v_B \cos \theta_B - \omega_{AB} r_{A/B} \cos \theta_{A/B}.$$

Note the negative sign is included because the x-component of $v_{A/B}$ is in the negative direction in Figure 14.8.

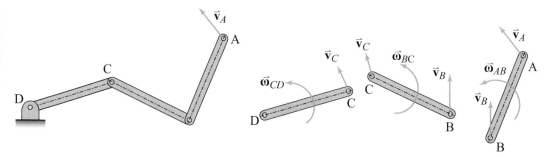

Figure 14.9: Mechanism consisting of three links, each with a velocity analysis.

y-dir:

$$(v_A)_y = (v_B)_y + (v_{A/B})_y$$
$$(v_A)_y = v_B \sin \theta_B + \omega_{AB} r_{A/B} \sin \theta_{A/B}.$$

With the components we have the solution:

$$\vec{\mathbf{v}}_A = (v_A)_x \,\hat{\imath} + (v_A)_y \,\hat{\jmath}$$
$$= \left[v_B \cos \theta_B - \omega_{AB} r_{A/B} \cos \theta_{A/B} \right] \hat{\imath} + \left[v_B \sin \theta_B + \omega_{AB} r_{A/B} \sin \theta_{A/B} \right] \hat{\jmath}.$$

The result agrees with that found using Vector Math. This generic demonstration solved for the velocity components of Point A, but we might have other scenarios where we'll know the velocity of both Points A and B and instead need to find the angular velocity.

Don't attempt to memorize this result; instead think through each problem using the diagram. The signs and the angles will change depending on the scenario of each problem, so "plugging and chugging" isn't possible. Compared to vector math, which follows a methodical procedure producing the correct result regardless of scenario, using the diagram and vector components requires more thought. We might consider this a weakness of the approach or a strength. It's a weakness in that we have to think more, but it's a strength because we gain insight into what's actually occurring in the movement of the object.

14.3 MULTIPLE RIGID BODY PARTS (MECHANISMS)

Many rigid body problems are "mechanisms" made up of multiple parts. As mentioned in Section 13.7, when these parts are connected by pins they have the same velocities (and acclerations). We use this to reason out the motion of the mechanism by doing multiple velocity analyses. Figure 14.9 shows a generic mechanism and the links separated for a velocity analysis of each.

We can march through the linkages, performing a velocity analysis of each. Here we'll use the vector math approach:

$$\vec{\mathbf{v}}_A = \vec{\mathbf{v}}_B + \vec{\omega}_{AB} \times \vec{\mathbf{r}}_{A/B}$$

Figure 14.10: Newtdog on penny-farthing (© E. Diehl).

$$\vec{v}_B = \vec{v}_C + \vec{\omega}_{BC} \times \vec{r}_{B/C}$$

$$\vec{v}_C = \vec{\omega}_{CD} \times \vec{r}_{C/D}$$

$$\vec{v}_A = \vec{\omega}_{CD} \times \vec{r}_{C/D} + \vec{\omega}_{BC} \times \vec{r}_{B/C} + \vec{\omega}_{AB} \times \vec{r}_{A/B}.$$

In Examples 14.3 and 14.4 we'll use this concept to work through the desired results.

Example 14.1
Newtdog is riding a penny-farthing bicycle (200 years before it was invented) which has a $d = 4$ ft front wheel. The bicycle is traveling at $v_A = 4$ ft/s, and the front wheel, which is slipping, is rotating at $\omega_{AB} = 3$ rad/s clockwise. Find the velocity of Point B on the wheel.

First, we should remember that if the wheel didn't slip Point B's velocity would be zero. We will use the component Velocity Diagram approach in this simple problem. Figure 14.11 presents the Velocity Diagram and assumes the velocity at B is in the positive direction. This is a good default assumption if this velocity is the desired the result. As we draw the diagram we can see that v_B will be less than v_A and possibly in the opposite direction because of the influence of $v_{B/A}$.

We write out the equation and then break it into components based on direction:

$$\vec{v}_B = \vec{v}_A + \vec{v}_{B/A}.$$

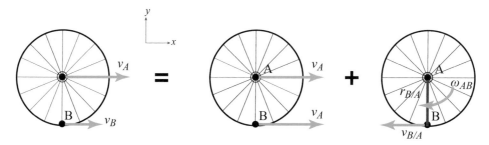

Figure 14.11: Velocity diagram from Example 14.1.

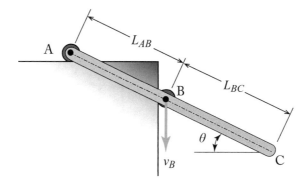

Figure 14.12: Example 14.2.

x-dir (note we designate signs based on the direction shown in the velocity diagram):

$$v_B = v_A - \omega_{AB} r_{B/A} = (4) - (3)(2) = -2 \text{ ft/s}.$$

The result shows the velocity at B is actually to the left which was expected, but we treated it as positive to demonstrate the default convention we should apply if we are unsure of the result:

$$\boxed{\vec{v}_B = 2.00 \text{ ft/s} \leftarrow}.$$

The next example is much more complicated and demonstrates all three velocity analysis methods covered in this class.

Example 14.2
Link ABC shown in Figure 14.12 has wheels at Points A and B that remain in contact on the horizontal and vertical surfaces. The dimensions are $L_{AB} = 4$ ft and $L_{BC} = 6$ ft. When $\theta = 25°$ the velocity of wheel B is $v_B = 1.5$ ft/s downward. Determine the velocity Point C.

We will use all three methods to solve this problem.

Vector Math:

We use a strategy to collect what we know (the velocity at B and the constrained direction of Point A) to find the angular velocity. Then we can use the angular velocity to find the velocity of Point C. This is a typical approach:

$$\vec{v}_A = \vec{v}_B + \vec{v}_{A/B}$$

$$\vec{v}_A = \vec{v}_B + \vec{\omega}_{ABC} \times \vec{r}_{A/B}$$

$$\vec{v}_A = v_A\hat{i} \quad \vec{v}_B = -v_B\hat{j} \quad \vec{\omega}_{ABC} = -\omega_{ABC}\hat{k} \quad \vec{r}_{A/B} = -L_{AB}\cos\theta\hat{i} + L_{AB}\sin\theta\hat{j}$$

$$v_A\hat{i} = -v_B\hat{j} - \omega_{ABC}\hat{k} \times [-L_{AB}\cos\theta\hat{i} + L_{AB}\sin\theta\hat{j}]$$

$$v_A\hat{i} = -v_B\hat{j} + \omega_{ABC}L_{AB}\cos\theta \underbrace{\hat{k}\times\hat{i}}_{\hat{j}} + \omega_{ABD}L_{AB}\sin\theta \underbrace{\hat{k}\times\hat{j}}_{-\hat{i}}$$

$$v_A\hat{i} = -v_B\hat{j} - \omega_{ABC}L_{AB}\sin\theta\hat{i} + \omega_{ABC}L_{AB}\cos\theta\hat{j}.$$

Collect the unit vector terms:

$$\hat{j}: \quad 0 = -v_B + \omega_{ABC}L_{AB}\cos\theta \quad (1.5) = \omega_{ABC}(4)\cos(25°) \quad \omega_{ABC} = 0.4138 \text{ rad/s}$$

$$\hat{i}: \quad v_A = \omega_{ABC}L_{AB}\sin\theta \quad v_A = (0.4138)(4)\sin(25°) \quad v_A = 0.6995 \text{ ft/s}.$$

We didn't need to find the velocity of A, but it's worth checking the result along the way to compare to the velocity at B:

$$\vec{v}_C = \vec{v}_B + \vec{v}_{C/B}$$

$$\vec{v}_C = (v_C)_x\hat{i} + (v_C)_y\hat{j} \quad \vec{v}_B = -v_B\hat{j} \quad \vec{\omega}_{ABC} = -\omega_{ABC}\hat{k} \quad \vec{r}_{C/B} = L_{BC}\cos\theta\hat{i} - L_{BC}\sin\theta\hat{j}$$

$$(v_C)_x\hat{i} + (v_C)_y\hat{j} = -v_B\hat{j} - \omega_{ABC}\hat{k} \times [L_{BC}\cos\theta\hat{i} - L_{BC}\sin\theta\hat{j}]$$

$$(v_C)_x\hat{i} + (v_C)_y\hat{j} = -v_B\hat{j} - \omega_{ABC}L_{BC}\cos\theta \underbrace{\hat{k}\times\hat{i}}_{\hat{j}} + \omega_{ABC}L_{BC}\sin\theta \underbrace{\hat{k}\times\hat{j}}_{-\hat{i}}$$

$$(v_C)_x\hat{i} + (v_C)_y\hat{j} = -v_B\hat{j} - \omega_{ABC}L_{BC}\sin\theta\hat{i} - \omega_{ABC}L_{BC}\cos\theta\hat{j}$$

$$\hat{i}: \quad (v_C)_x = -\omega_{ABC}L_{BC}\sin\theta \quad (v_C)_x = -(0.4138)(6)\sin(25°)$$
$$(v_C)_x = -1.049 \text{ ft/s}$$

$$\hat{j}: \quad (v_C)_y = -v_B - \omega_{ABC}L_{BC}\cos\theta \quad (v_C)_y = -(1.5) - (0.4138)(6)\cos(25°)$$
$$(v_C)_y = -3.750 \text{ ft/s}$$

$$\vec{v}_C = (-1.049)\hat{i} + (-3.750)\hat{j} \text{ ft/s}$$

$$|\vec{v}_C| = \sqrt{(-1.049)^2 + (-3.750)^2} = 3.894 \text{ ft/s} \quad \theta = \tan^{-1}\left(\frac{3.750}{1.049}\right) = 74.37°$$

$$\vec{v}_C = 3.894 \text{ ft/s} \diagup 74.37°.$$

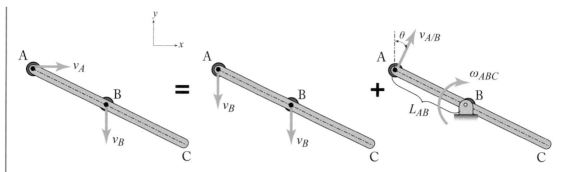

Figure 14.13: Velocity diagram for Point A in Example 14.2.

Vector Components of Velocity Diagram:
We draw the velocity diagram in Figure 14.13 to solve for the velocity of Point A. As we saw in the Vector Math approach, we're actually trying to find the angular velocity of the bar. We use B as the reference since we know this value. The velocity of A must be horizontal, so when deciding the direction of rotation, the rotation must have an upward vertical sense in order to remove the downward vertical component of B. This is an important step, so re-read this statement carefully:

$$\vec{\mathbf{v}}_A = \vec{\mathbf{v}}_B + \vec{\mathbf{v}}_{A/B}.$$

We break this into components with signs that correspond to the diagram.

y-dir:

$$0 = -v_B + v_{A/B} \cos \theta \qquad 0 = -v_B + \omega_{ABC} L_{AB} \cos \theta$$
$$0 = -(1.5) + \omega_{ABC} (4) \cos(25°) \qquad \omega_{ABC} = 0.4138 \text{ rad/s } \circlearrowleft .$$

Be sure to recognize the strategy of using the known direction of A to get the angular velocity in the calculation step. We don't need to find the x-direction, but will anyway for completeness.

x-dir:

$$v_A = 0 + v_{A/B} \sin \theta \qquad v_A = 0 + \omega_{ABD} L_{AB} \sin \theta$$
$$v_A = 0 + (0.4138)(4) \sin(25°) = 0.6995 \text{ ft/s}.$$

We use B as the reference again, and will re-draw the Velocity Diagram in Figure 14.14. We could try to use the same diagram for both, but doing so might make it too cluttered and confusing. We draw both velocity components of C in the positive direction even though we're pretty sure they will be negative.

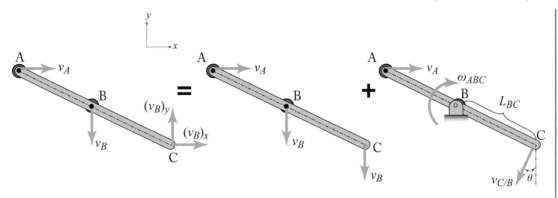

Figure 14.14: Velocity diagram for Point C in Example 14.2.

$$\vec{\mathbf{v}}_C = \vec{\mathbf{v}}_B + \vec{\mathbf{v}}_{C/B}.$$

x-dir:

$$(v_C)_x = (0) - v_{C/B} \sin\theta \quad (v_C)_x = (0) - \omega_{ABC} L_{BC} \sin\theta$$
$$(v_C)_x = (0) - (0.4138)(6)\sin(25°) = -1.049 \text{ ft/s}.$$

y-dir:

$$(v_C)_y = v_B - v_{C/B} \cos\theta \quad (v_C)_y = v_B - \omega_{ABC} L_{BC} \cos\theta$$
$$(v_C)_y = -(1.5) - (0.4138)(6)\cos(25°) = -3.750 \text{ ft/s}.$$

This is the same result as with Vector Math.

Vector Triangles:
We use the Velocity Diagrams in Figures 14.13 and 14.14 to construct vector triangles in Figure 14.15.

The first vector triangle of Figure 14.15 represents:

$$\vec{\mathbf{v}}_A = \vec{\mathbf{v}}_B + \vec{\mathbf{v}}_{A/B}.$$

We can use right triangle geometry to find the magnitudes of $\vec{\mathbf{v}}_A$ and $\vec{\mathbf{v}}_{A/B}$

$$v_{A/B} = \frac{v_B}{\cos\theta} = \frac{(1.5)}{\cos(25°)} = 1.655 \text{ ft/s}$$

$$v_A = v_B \tan\theta = (1.5)\tan(25°) = 0.6995 \text{ ft/s} \quad \text{(not needed)}$$

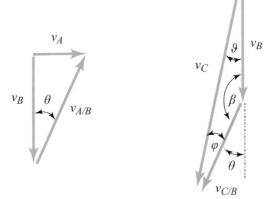

Figure 14.15: Vector triangles for Example 14.2.

$$v_{A/B} = \omega_{ABC} L_{AB} \quad \omega_{ABC} = \frac{v_{A/B}}{L_{AB}} = \frac{(1.655)}{(4)} = 0.4138 \text{ rad/s}.$$

The second vector triangle of Figure 14.15 respresents:

$$\vec{v}_C = \vec{v}_B + \vec{v}_{C/B}.$$

The magnitude of $\vec{v}_{C/B}$ is found from:

$$v_{C/B} = \omega_{ABC} L_{BC} = (0.4138)(6) = 2.483 \text{ ft/s}.$$

We can find the obtuse angle of the triangle from the supplement of θ:

$$\beta = 180° - \theta = 180° - 25° = 155°.$$

We know two adjacent sides and the angle in between so we use LoC to find the magnitude of $\vec{v}_C$:

$$v_C = \sqrt{v_B{}^2 + v_{C/B}{}^2 - 2 v_B v_{C/B} \cos \beta}$$

$$= \sqrt{(1.5)^2 + (2.483)^2 - 2(1.5)(2.483)\cos(155°)} = 3.894 \text{ ft/s}.$$

We can find the direction using LoS:

$$\frac{v_{C/B}}{\sin \vartheta} = \frac{v_C}{\sin \beta} \quad \vartheta = \sin^{-1}\left[\frac{v_{C/B}}{v_C} \sin \beta\right] = \sin^{-1}\left[\frac{(2.483)}{(3.894)} \sin(155°)\right] = 15.63°$$

subtract from 90° for angle w.r.t. horiz.

Answer: $\boxed{\vec{v}_C = 3.89 \text{ ft/s} \nearrow 74.4°}$.

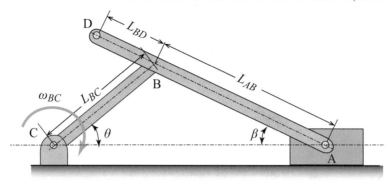

Figure 14.16: Example 14.3.

The answer has been confirmed in three different approaches. The vector triangle method is an especially useful method to confirm the results graphically using a careful sketch with a straight edge on graph paper. Many textbooks use the vector math approach almost exclusively. While vector math is an essential language for engineering communication, we can sometimes lose sight of what is actually taking place until we reach the result.

We will use the Vector Components of Velocity Diagram method because it helps provide a sense of what is taking place and is especially good for visualizing the motion.

Example 14.3

A "slider-crank" mechanism, as shown in Figure 14.16, is one of the basic machines commonly studied in Dynamics and includes reciprocating engines. Crank BC rotates at $\omega_{BC} = 8$ rad/s clockwise and is in the position $\theta = 40°$. The dimensions are $L_{AB} = 500$ mm, $L_{BC} = 300$ mm, and $L_{BD} = 150$ mm. Determine the velocity of Point D.

This example demonstrates a mechanism consisting of multiple parts generating multiple equations that allow us to find the unknowns needed to get the final answer.

We start with finding the unknown angle β using LoS:

$$\frac{L_{AB}}{\sin \theta} = \frac{L_{BC}}{\sin \beta} \quad \beta = \sin^{-1}\left[\frac{L_{BC}}{L_{AB}} \sin \theta\right] = \sin^{-1}\left[\frac{(300)}{(500)} \sin\left(40°\right)\right] = 22.69°.$$

The Velocity Diagram of crank BC in Figure 14.17 only requires one drawing since it is rotating about a fixed point.

$$\vec{v}_B = \vec{v}_C + \vec{v}_{B/C} = (0) + \vec{v}_{B/C}.$$

The magnitude of $\vec{v}_B$ is found from : $v_B = v_{B/C} = \omega_{BC} L_{BC} = (0.300)(8) = 2.400$ m/s.
The direction is known from the geometry of Figure 14.17.

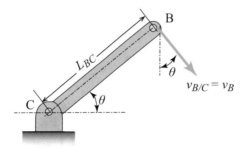

Figure 14.17: Velocity diagram of link BC of Example 14.3.

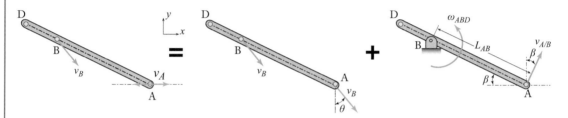

Figure 14.18: Velocity diagram of rod ABC in Example 14.3 with focus on Point A.

The Velocity Diagram of rod ABD is done in two stages to avoid confusion. We start with a focus on Point A in Figure 14.18. Note that we show A as moving either to the left or the right. In this instance, we're pretty sure it's moving to the right, but in some mechanisms we won't be able to know for sure, only that it will be traveling along a particular line due to geometry constraints:

$$\vec{v}_A = \vec{v}_B + \vec{v}_{A/B}.$$

y-dir:
$$0 = -v_B \cos\theta + v_{A/B} \cos\beta \qquad 0 = -v_B \cos\theta + \omega_{ABD} L_{AB} \cos\beta$$

$$0 = -(2.400) \cos\left(40°\right) + \omega_{ABD}(0.500) \cos\left(22.69°\right) \qquad \omega_{ABD} = 3.985 \text{ rad/s } \circlearrowleft .$$

We don't need to use the x-direction because we weren't asked for the velocity at A, but let's find it anyway.

x-dir:
$$v_A = v_B \sin\theta + v_{A/B} \sin\beta \qquad v_A = v_B \sin\theta + \omega_{ABD} L_{AB} \sin\beta$$

$$v_A = (2.400) \sin\left(40°\right) + (3.985)(0.500) \sin(22.69°) = 2.311 \text{ m/s } \rightarrow .$$

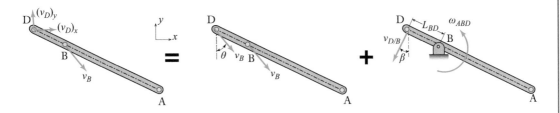

Figure 14.19: Velocity diagram of rod *ABD* in Example 14.3 with focus on Point *D*.

We redraw a Velocity Diagram for rod *ABD* in Figure 14.19, this time focusing on Point *D*. We draw the components at *D* in the positive direction by default:

$$\vec{v}_D = \vec{v}_B + \vec{v}_{D/B}.$$

x-dir:
$$(v_D)_x = v_B \sin\theta - v_{D/B}\sin\beta \qquad (v_D)_x = v_B \sin\theta - \omega_{ABD}L_{BD}\sin\beta$$

$$(v_D)_x = (2.400)\sin\left(40°\right) - (3.985)(0.150)\sin\left(22.69°\right) = 1.312 \text{ m/s.}$$

y-dir:
$$(v_D)_y = -v_B \cos\theta - v_{D/B}\cos\beta \qquad (v_D)_y = -v_B \cos\theta - \omega_{ABD}L_{BD}\cos\beta$$

$$(v_D)_y = -(2.400)\cos\left(40°\right) - (3.985)(0.150)\cos\left(22.69°\right) = -2.390 \text{ m/s}$$

$$\vec{v}_D = (1.312)\hat{\imath} + (-2.390)\hat{\jmath} \text{ m/s}$$

$$\left|\vec{v}_D\right| = \sqrt{(1.312)^2 + (-2.390)^2} = 2.726 \text{ m/s} \quad \theta = \tan^{-1}\left(\frac{2.390}{1.312}\right) = 61.24°$$

$$\vec{v}_D = 2.726 \text{ m/s} \searrow 61.24°.$$

Answer: $\boxed{\vec{v}_D = 2.73 \text{ m/s} \searrow 61.2°}$.

Example 14.4
A "four-bar mechanism" is also among the most basic of basic mechanisms and consists of three links plus the distance between the fixed pivot joints called the ground link (which is treated like a stationary fourth bar). Figure 14.20 presents a typical four-bar mechanism. In the position shown, bar *CD* has an angular velocity of $\omega_{CD} = 12$ rad/s counter-clockwise as if it were being

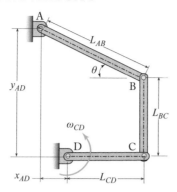

Figure 14.20: **Example** 14.4.

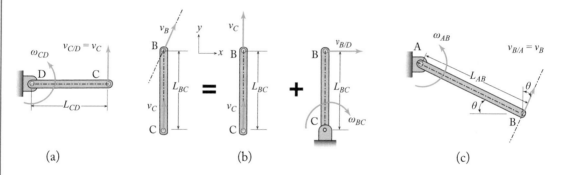

Figure 14.21: Velocity diagrams of Example 14.4.

driven by a motor we don't see. The dimensions are L_{AB} = 600 mm, L_{BC} = 400 mm, L_{CD} = 350 mm, and x_{AD} = 200 mm. Determine the velocity of B.

A typical mechanism analysis begins with position analysis to determine dimensions or angles not provided:

$$y_{AD} = L_{BC} + \sqrt{L_{AB}^2 - (L_{CD} + x_{AD})^2}$$

$$= (400) + \sqrt{(600)^2 - ((350) + (200))^2} = 639.9 \text{ mm}$$

$$\theta = \sin^{-1}\left(\frac{y_{AD} - L_{BC}}{L_{AB}}\right) = \sin^{-1}\left(\frac{(639.9) - (400)}{(600)}\right) = 23.56°.$$

We separate the bars of the mechanism in Figure 14.21 to create three different Velocity Diagrams for analysis.

Links CD and AB are both in pure rotation. The direction of C is upward as shown in Figure 14.21a, and we can easily find the velocity:

$$v_C = v_{C/D} = \omega_{CD} L_{CD} = (12)(0.350) = 4.200 \text{ m/s}.$$

We skip over to Link AB in Figure 14.21c and see the velocity of B must be perpendincular to it. These two constraints (C must be upwards and B must follow the line perpendicular to AB) help determine the rotation of link BC. As shown in Figure 14.21b, we assume it rotates clockwise. Doing an analysis of BC we see the velocity component of B in the y-direction is entirely due to velocity of C:

$$\vec{\mathbf{v}}_B = \vec{\mathbf{v}}_C + \vec{\mathbf{v}}_{B/C}.$$

x-dir:

$$(v_B)_x = (0) + v_{B/C} = (0) + \omega_{BC} L_{BC}.$$

y-dir:

$$(v_B)_y = v_C + (0) = 4.200 \text{ m/s}.$$

For Link AB:

$$\vec{\mathbf{v}}_B = \vec{\mathbf{v}}_{A/B}.$$

y-dir:

$$(v_B)_y = v_{A/B} \cos \theta = \omega_{AB} L_{AB} \cos \theta = \omega_{AB} (0.600) \cos (23.56°) = 4.200 \text{ m/s}$$

$$\omega_{AB} = 7.637 \text{ rad/s} \circlearrowleft .$$

x-dir:

$$(v_B)_x = v_{A/B} \sin \theta = \omega_{AB} L_{AB} \sin \theta = (7.637)(0.600) \sin (23.56°) = 1.831 \text{ m/s}.$$

With this we can also find the angular velocity of BC:

$$(v_B)_x = \omega_{BC} L_{BC} = \omega_{BC} (0.400) = 1.831 \text{ m/s}.$$

Even though we weren't asked for the angular velocity of link BC, it's worth finding it now since this example problem will also be used in Classes 15 and 16:

$$\omega_{BC} = 4.578 \text{ rad/s} \circlearrowright .$$

With the components of B known, we can find the velocity:

$$\vec{v}_B = (1.831)\hat{\imath} + (4.200)\hat{\jmath} \text{ m/s}$$

$$\left|\vec{v}_B\right| = \sqrt{(1.831)^2 + (4.200)^2} = 4.582 \text{ m/s}$$

$$\theta = \tan^{-1}\left(\frac{4.200}{1.831}\right) = 66.45° \quad (\text{also } 90° - \theta)$$

$$\vec{v}_B = 4.582 \text{ m/s} \nearrow 66.45°.$$

Answer: $\boxed{\vec{v}_B = 4.58 \text{ m/s} \nearrow 66.5°}$.

Book 2 - Class 15

https://www.youtube.com/watch?v=x1wSLGAL2k8

CLASS 15

Velocity Analysis Using the Instantaneous Center of Rotation

B.L.U.F. (Bottom Line Up Front)

- At any instant there exists an imaginary point for a rigid body called the "Instantaneous Center of Rotation" (ICR).

- The velocity of the ICR is zero and the rigid body appears to be in pure rotation about this point.

- The ICR is found at the intersection of lines drawn perpendicular to the known velocity directions.

- The ICR doesn't have to be on the body.

- Once the ICR location is identified, and the angular speed calculated, any velocity on the body can be found.

15.1 INSTANTANEOUS CENTER OF ROTATION

Velocity Analysis of two-dimensional Rigid Body Motion can be greatly simplified by identifying the "Instantaneous Center of Rotation" (ICR). The ICR is an imaginary point in the plane of rotation that the rigid body appears to be rotating about as if it were a fixed pivot joint. Even when an object is translating and rotating, taking a snap-shot of the motion (thus "instantaneous") we can find a magical fixed point where we can treat the object as pure rotation. Figure 15.1 shows a generic object with translation and rotation next to the equivalent pure rotation about an ICR.

The general equation for Velocity Analysis is still valid except that the ICR is a fixed point:

$$\vec{v}_A = \vec{v}_{ICR} + \vec{v}_{A/ICR} = (0) + \vec{\omega}_A \times \vec{r}_{A/ICR}.$$

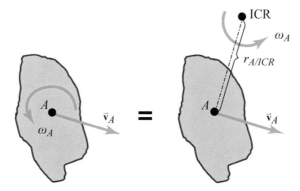

Figure 15.1: Translation and rotation equivalent as pure rotation about ICR.

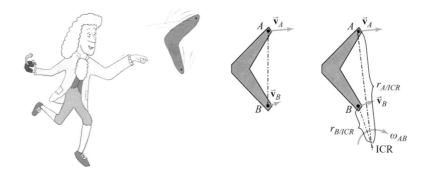

Figure 15.2: Newtdog's boomerang appears to be in pure rotation about and ICR (© E. Diehl).

The ICR is located on a line perpendicular to the direction of the velocity at any point on the object. The distance along this line is found from the scalar version of the pure rotation velocity equation: $v_A = \omega_A r_{A/ICR}$.

Figure 15.2 shows Newtdog throwing the boomerang again. The labeled diagrams repeat the velocity vectors on two points. The diagram to the far right has lines drawn perpendicular to both velocity directions. The intersection of these lines is the location of the ICR.

15.2   LOCATING THE INSTANTANEOUS CENTER OF ROTATION

The ICR is identified differently depending on the scenario. The position vector(s), $\vec{\mathbf{r}}_{point/ICR}$, from the ICR to each point is perpendicular to the velocity at that point. The ICR may or may not be on the body. Below are three general cases.

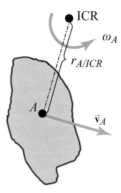

Figure 15.3: ICR determined from known translation and rotation.

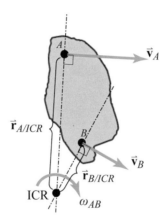

Figure 15.4: ICR determined from known velocities of two points.

1. Figure 15.3 shows when the velocity of a point, $\vec{v}_A$, and the rotation, $\vec{\omega}_A$, are known. The magnitude of $r_{A/\text{ICR}}$ is found from $r_{A/\text{ICR}} = \frac{v_A}{\omega_A}$.

2. Figure 15.4 shows when the velocity at two points, $\vec{v}_A$ and $\vec{v}_B$, are different directions and their magnitudes are known. The ICR is found geometrically from where the perpendicular lines cross.

3. Figure 15.5 shows when the velocity at two points, $\vec{v}_A$ and $\vec{v}_B$, are the same direction (parallel), but have different magnitudes which are known. We can see that the ICR can be either away from or on the body in these diagrams.

Figure 15.6 presents some drawings of examples applying ICR to parts and mechanisms. Each rigid body has its own ICR.

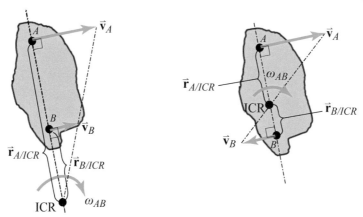

Figure 15.5: ICR found from two known velocities whose perpendicular line align.

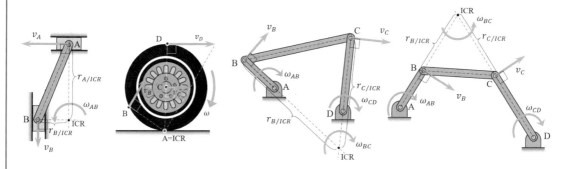

Figure 15.6: ICR of a (a) double sliding link, (b) a non-slipping wheel, (c) a four-bar mechanism, and (d) another four-bar mechanism.

The double sliding link on the left has an ICR perpendicular to the paths. Note that a common mistake is to take it at the intersection of the paths. The ICR of the non-slipping wheel is the point in contact with the ground. The ICR of the middle link in a four-bar mechanism (called the "coupler") is at the intersection of lines extending from the other two links (called the "crank" and "rocker") which are both in pure rotation.

15.3   USING GEOMETRY TO FIND THE INSTANTANEOUS CENTER OF ROTATION

While the ICR velocity method is often an easy alternative to one of the other Velocity Analysis methods presented in Class 14, the geometry needed to locate the ICR can sometimes be challenging. Figure 15.7 illustrates a generic ICR problem where the distances $r_{A/\text{ICR}}$ and $r_{B/\text{ICR}}$ are needed.

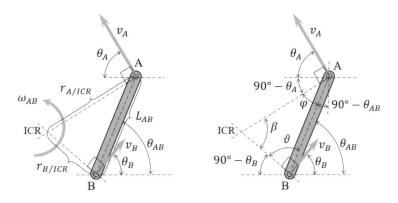

Figure 15.7: Using geometry to find the ICR.

Here we assume that we know the length of the link (L_{AB}) and the angles of velocities (θ_A and θ_B) and link (θ_{AB}). We would generally know the angles of the velocities based on the arrangement of other parts. To find the distances $r_{A/ICR}$ and $r_{B/ICR}$ we need the interior angles of the triangle formed with the link length. By looking for supplementary and complementary angles in the arrangement, as well as applying other geometry identities, we can see $\vartheta = 90° - \theta_{AB} + \theta_B$, $\varphi = \theta_A + \theta_{AB} - 90°$, and $\beta = 180° - \theta_A - \theta_B$. We can use LoS with these angles to find the distances. Note that these equations are not universal and need to be reasoned for each scenario.

ICR works with velocity analysis, but is not suitable with acceleration analysis as we will see in the next class.

The same example problems from Chapter 14 are presented here, but this time they are solved using ICR.

Example 15.1   (repeat of Example 14.1)
Newtdog is riding a penny-farthing bicycle in Figure 15.8 which has a $d = 4$ ft front wheel. The bicycle is traveling at $v_A = 4$ ft/s, and the front wheel, which is slipping, is rotating at $\omega_{AB} = 3$ rad/s clockwise. Find the velocity of Point B on the wheel.

We draw the wheel as an ICR diagram in Figure 15.9. Since the velocities are parallel the ICR lands on the line that runs through them. We find the location on that line by dividing the known velocity of A by the angular velocity:

$$r_{A/ICR} = \frac{v_A}{\omega_{AB}} = \frac{(4)}{(3)} = 1.333 \text{ ft}$$

$$r_{B/ICR} = \frac{d}{2} - r_{A/ICR} = \frac{(4)}{2} - (1.333) = 0.6667 \text{ ft}$$

Figure 15.8: Newtdog on a penny-farthing (repeat of Figure 14.10) (© E. Diehl).

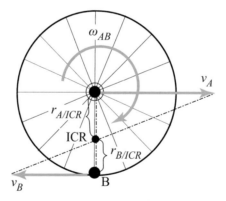

Figure 15.9: ICR diagram for Example 15.1.

$$v_B = \omega_{AB} r_{B/ICR} = (3)(0.6667) = 2.000 \text{ ft/s} \longleftarrow$$

$$\boxed{\vec{v}_B = 2.00 \text{ ft/s} \leftarrow}.$$

This is the same answer we found in Example 14.1. It is interesting that the ICR is between these parallel velocities on the line perpendicular to both. The ICR method is often useful as a check of the results from the other velocity analysis methods.

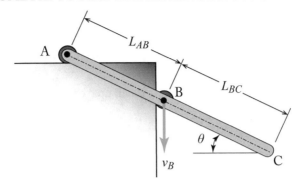

Figure 15.10: Example 15.2 (repeat of Figure 14.12).

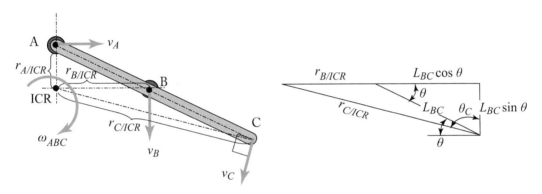

Figure 15.11: ICR diagram and side sketch of Example 15.2.

Example 15.2   (Repeat of Example 14.2)

Link *ABC* shown in Figure 15.10 has wheels at Points *A* and *B* that remain in contact on the horizontal and vertical surfaces, respectively. The dimensions are $L_{AB} = 4$ ft and $L_{BC} = 6$ ft. When $\theta = 25°$ the velocity of wheel *B* is $v_B = 1.5$ ft/s downward. Determine the velocity Point *C*.

We draw the ICR diagram in Figure 15.11 along with a side sketch of an interior triangle to help us find the distances we need. It is good practice to make side sketches like this when too much detail might clutter up the main diagram.

From the ICR diagram on the left we find the lengths we need for Points *A* and *B*:

$$r_{A/\text{ICR}} = L_{AB} \sin \theta = (4) \sin \left(25°\right) = 1.690 \text{ ft}$$

$$r_{B/\text{ICR}} = L_{AB} \cos \theta = (4) \cos \left(25°\right) = 3.625 \text{ ft}.$$

The angular velocity of the link is found from:

$$\omega_{ABC} = \frac{v_B}{r_{B/ICR}} = \frac{(1.5)}{(3.625)} = 0.4138 \text{ rad/s } \circlearrowleft$$

From the side sketch on the right we can break apart the geometry to find the ICR for Point C:

$$r_{C/ICR} = \sqrt{\left(r_{B/ICR} + L_{BC}\cos\theta\right)^2 + (L_{BC}\sin\theta)^2}$$

$$= \sqrt{((3.625) + (6)\cos(25°))^2 + ((6)\sin(25°))^2} = 9.411 \text{ ft}$$

$$v_C = \omega_{ABC}r_{C/ICR} = (0.4138)(9.411) = 3.894 \text{ ft/s.}$$

Not asked for, but we can find the velocity at Point A easily:

$$v_A = \omega_{ABC}r_{A/ICR} = (0.4138)(1.690) = 0.6995 \text{ ft/s.}$$

The angle of the result can be found from the side sketch:

$$\theta_C = \tan^{-1}\left[\frac{r_{B/ICR} + L_{BC}\cos\theta}{L_{BC}\sin\theta}\right] = \tan^{-1}\left[\frac{((3.625) + (6)\cos(25°))}{((6)\sin(25°))}\right] = 74.37°.$$

Answer: $\boxed{\vec{v}_C = 3.89 \text{ ft/s} \diagup 74.4°}$ this matches the results from Example 14.2. We've found this same solution four different ways. Go back and compare the methods used in Example 14.2 to this and decide for yourself which was easiest.

Example 15.3   (repeat of Example 14.3)

Crank BC of the slider-crank in Figure 15.12 rotates at $\omega_{BC} = 8$ rad/s clockwise and is in a position where $\theta = 40°$. The dimensions are $L_{AB} = 500$ mm, $L_{BC} = 300$ mm, and $L_{BD} = 150$ mm. Determine the velocity of Point D.

We found $\beta = 22.69°$ in Example 14.3. In Figure 15.13 we draw the two links of the slider crank. Link BC is in pure rotation and the magnitude of $\vec{v}_B$ is found to be:

$$v_B = v_{B/C} = \omega_{BC}L_{BC} = (0.300)(8) = 2.400 \text{ m/s.}$$

The geometry of the ICR lengths for link ABD can be found by breaking the triangles into pieces, introducing Points E and F:

$$L_{AE} = L_{AB}\sin\beta = (0.500)\sin(22.69°) = 0.1929 \text{ m}$$

$$L_{BE} = L_{AB}\cos\beta = (0.500)\cos(22.69°) = 0.4613 \text{ m}$$

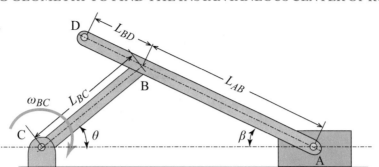

Figure 15.12: **Example** 15.3 (repeat of Figure 14.16).

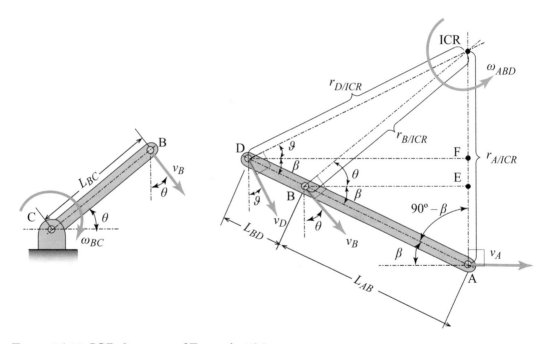

Figure 15.13: **ICR** diagrams of Example 15.3.

$$L_{DF} = (L_{AB} + L_{BD}) \cos \beta = ((0.500) + (0.150)) \cos (22.69°) = 0.5997 \text{ m}$$

$$L_{AF} = (L_{AB} + L_{BD}) \sin \beta = ((0.500) + (0.150)) \sin (22.69°) = 0.2508 \text{ m}$$

$$r_{A/ICR} = L_{AE} + L_{BE} \tan \theta = (0.1929) + (0.4613) \tan (40°) = 0.5800 \text{ m}$$

$$r_{B/ICR} = \frac{L_{AB} \cos \beta}{\cos \theta} = \frac{(0.500) \cos (22.69°)}{\cos (40°)} = 0.6022 \text{ m}$$

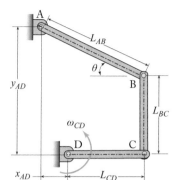

Figure 15.14: Example 15.4 (repeat of Figure 14.20).

$$r_{D/\text{ICR}} = \sqrt{L_{DF}^2 + \left(r_{A/\text{ICR}} - L_{AF}\right)^2}$$
$$= \sqrt{(0.5997)^2 + ((0.5800) - (0.2508))^2} = 0.6841 \text{ m}$$

$$\vartheta = \cos^{-1}\frac{L_{DF}}{r_{D/\text{ICR}}} = \cos^{-1}\left(\frac{0.5997}{0.6841}\right) = 28.76°.$$

We can find the velocities with these ICR distances:

$$\omega_{ABD} = \frac{v_B}{r_{B/\text{ICR}}} = \frac{(2.400)}{(0.6022)} = 3.985 \text{ rad/s} \ \circlearrowleft$$

$$v_A = \omega_{ABD}r_{A/\text{ICR}} = (3.985)(0.5800) = 2.312 \text{ m/s} \rightarrow$$
$$v_D = \omega_{ABD}r_{D/\text{ICR}} = (3.985)(0.6841) = 2.726 \text{ m/s} \searrow .$$

To report the angle with respect to the horizontal axis:

$$90° - \vartheta = 90° - 28.76° = 61.24°.$$

Answer: $\boxed{\vec{v}_D = 2.73 \text{ m/s} \searrow 61.2°}$ which matches what was found in Example 14.3.

Example 15.4 (repeat of Example 14.4)
In the position shown, bar CD of the four-bar mechanism in Figure 15.14 has an angular velocity of $\omega_{CD} = 12$ rad/s counter-clockwise. The dimensions are $L_{AB} = 600$ mm, $L_{BC} = 400$ mm, $L_{CD} = 350$ mm, and $x_{AD} = 200$ mm. Determine the velocity of Point B.

We draw the ICR diagram of the entire mechanism in Figure 15.15 to emphasize how the lines extended from the crank and rocker form the ICR of the coupler. We see this simplifies recognizing the geometry.

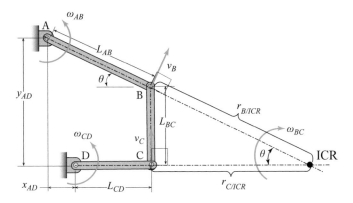

Figure 15.15: ICR diagram of Example 15.4.

We find the magnitude of $\vec{v}_C$:

$$v_C = \omega_{CD} L_{CD} = (0.350)(12) = 4.200 \text{ m/s}.$$

The ICR geometry (knowing from Example 14.4 that $\theta = 23.56°$):

$$r_{C/ICR} = \frac{L_{BC}}{\tan \theta} = \frac{(0.400)}{\tan (23.56°)} = 0.9173 \text{ m}$$

$$r_{B/ICR} = \frac{L_{BC}}{\sin \theta} = \frac{(0.400)}{\sin (23.56°)} = 1.001 \text{ m}.$$

The angular velocity of the coupler (Link BC) is:

$$\omega_{BC} = \frac{v_C}{r_{C/ICR}} = \frac{(4.200)}{(0.9173)} = 4.579 \text{ rad/s} \circlearrowleft .$$

The velocity of B:

$$v_B = \omega_{BC} r_{B/ICR} = (4.579)(1.001) = 4.582 \text{ m/s}.$$

We note that the direction is

$$90° - \theta = 90° - (23.56°) = 66.44°.$$

Answer: $\boxed{\vec{v}_B = 4.58 \text{ m/s} \nearrow 66.5°}$ and this matches the answer in Example 14.4.

We see that performing a velocity analysis using ICR can greatly simplify the process as in Example 15.4, but also presents some challenging geometry as in Example 14.3.

Book 2 - Class 16

https://www.youtube.com/watch?v=s1DMG4lXLnw

CLASS 16

Acceleration Analysis (Part 1)

B.L.U.F. (Bottom Line Up Front)

- Acceleration Analysis is approached similarly to Velocity Analysis using relative motion: $\vec{a}_A = \vec{a}_B + \vec{a}_{A/B}$.

- Acceleration Analysis is more complicated than Velocity Analysis because the relative motion portion has two parts: normal and tangential contributions: $\vec{a}_A = \vec{a}_B + (\vec{a}_{A/B})_t + (\vec{a}_{A/B})_n$.

- A Velocity Analysis is usually needed before doing an Acceleration Analysis.

- It is difficult to use the Vector Triangle approach for Acceleration Analysis.

16.1   RELATIVE ACCELERATION ON A RIGID BODY

As with Particle Kinematics (Classes 4 (vol. 1) and 5 (vol. 1)), Acceleration Analysis of Rigid Body Motion is more complicated than Velocity Analysis due to extra terms. Just like Rigid Body Velocity Analysis, we use relative motion of two points on a rigid body to perform an Rigid Body Acceleration Analysis:

$$\boxed{\vec{a}_A = \vec{a}_B + \vec{a}_{A/B}}.$$

Unlike Velocity Analysis, the relative motion part of Acceleration Analysis can be further broken down into two parts, tangential and normal:

$$\vec{a}_A = \vec{a}_B + (\vec{a}_{A/B})_t + (\vec{a}_{A/B})_n.$$

In Figure 16.1, Newtdog and Wormy go down a hill inside a rigid wheel. If we use the acceleration of the center of the wheel (Point B) as the reference, we find the total acceleration at a point on the wheel (Point A) by adding the tangential and normal accelerations to it.

To find out why these two terms exist, we go back to basics and recall that two points on a rigid body can be described using the position vector:

$$\vec{r}_A = \vec{r}_B + \vec{r}_{A/B}.$$

Figure 16.1: Newtdog and Wormy in an accelerating wheel (© E. Diehl).

To find the velocity we take the time derivative of the position vector:

$$\frac{d}{dt}\left[\vec{r}_A = \vec{r}_B + \vec{r}_{A/B}\right] \implies \vec{v}_A = \vec{v}_B + \vec{v}_{A/B}$$
$$\implies \vec{v}_A = \vec{v}_B + \vec{\omega} \times \vec{r}_{A/B}.$$

Acceleration is the time derivative of the velocity vector and because of the product rule we've got an extra term:

$$\frac{d}{dt}\left[\vec{v}_A = \vec{v}_B + \vec{\omega} \times \vec{r}_{A/B}\right] \implies \vec{a}_A = \vec{a}_B + \frac{d}{dt}\left[\vec{\omega} \times \vec{r}_{A/B}\right]$$
$$\implies \vec{a}_A = \vec{a}_B + \vec{\alpha} \times \vec{r}_{A/B} + \vec{\omega} \times \vec{v}_{A/B}.$$

We already found that $\vec{v}_{A/B} = \vec{\omega} \times \vec{r}_{A/B}$, so we substitute and get:

$$\vec{a}_A = \vec{a}_B + \vec{\alpha} \times \vec{r}_{A/B} + \vec{\omega} \times \left(\vec{\omega} \times \vec{r}_{A/B}\right).$$

We see there are a total of four terms in rigid body acceleration equation rather than the three we have in the velocity equation. In two dimensions the rotation is only about an axis out of the plane, so $\vec{\omega} = \omega\hat{k}$ and $\vec{\alpha} = \alpha\hat{k}$. Since we are performing a cross product twice in the last term, it becomes negative, meaning that it points in the opposite direction of the relative position vector:

$$\boxed{\vec{a}_A = \vec{a}_B + \alpha\hat{k} \times \vec{r}_{A/B} - \omega^2 \vec{r}_{A/B}}.$$

This is the equation we will use when using vector math for Acceleration Analysis of two-dimensional rigid bodies based on a stationary frame of reference. The significance of this last statement will become clear in Class 18 when we re-introduce Coriolis acceleration.

16.2   SOLUTION METHOD USING VECTOR MATH

Below is a demonstration of solving a two-dimensional acceleration problem using vector math. We use generic vectors that are in the positive direction:

$$\vec{a}_A = (a_A)_x\hat{i} + (a_A)_y\,\hat{j}$$

$$\vec{a}_B = (a_B)_x\hat{i} + (a_B)_y\,\hat{j}$$

$$\vec{\omega}_{AB} = \omega_{AB}\hat{k}$$

$$\vec{\alpha}_{AB} = \alpha_{AB}\hat{k}$$

$$\vec{r}_{A/B} = (r_{A/B})_x\hat{i} + (r_{A/B})_y\,\hat{j}.$$

Applied to:

$$\vec{a}_A = \vec{a}_B + \alpha\hat{k} \times \vec{r}_{A/B} - \omega^2\vec{r}_{A/B}.$$

We get:

$$(a_A)_x\hat{i} + (a_A)_y\,\hat{j} = (a_B)_x\hat{i} + (a_B)_y\,\hat{j} + \alpha_{AB}\hat{k} \times \left((r_{A/B})_x\hat{i} + (r_{A/B})_y\,\hat{j}\right)$$
$$- \omega_{AB}^2\left((r_{A/B})_x\hat{i} + (r_{A/B})_y\,\hat{j}\right)$$

$$(a_A)_x\hat{i} + (a_A)_y\,\hat{j} = (a_B)_x\hat{i} + (a_B)_y\,\hat{j} + \alpha_{AB}\hat{k} \times (r_{A/B})_x\hat{i} + \alpha_{AB}\hat{k} \times (r_{A/B})_y\,\hat{j}$$
$$- \omega_{AB}^2\,(r_{A/B})_x\,\hat{i} - \omega_{AB}^2\,(r_{A/B})_y\,\hat{j}$$

$$(a_A)_x\hat{i} + (a_A)_y\,\hat{j} = (a_B)_x\hat{i} + (a_B)_y\,\hat{j} + \alpha_{AB}(r_{A/B})_x\,\underbrace{\hat{k} \times \hat{i}}_{+\,\hat{j}} + \alpha_{AB}(r_{A/B})_y\,\underbrace{\hat{k} \times \hat{j}}_{-\hat{i}}$$
$$- \omega_{AB}^2(r_{A/B})_x\hat{i} - \omega_{AB}^2\,(r_{A/B})_y\,\hat{j}$$

$$\vec{a}_A = \left[(a_B)_x - \alpha_{AB}(r_{A/B})_y - \omega_{AB}^2(r_{A/B})_x\right]\hat{i} + \left[(a_B)_y + \alpha_{AB}(r_{A/B})_x - \omega_{AB}^2(r_{A/B})_y\right]\hat{j}.$$

We can write the components as:

$$(a_A)_x = (a_B)_x - \alpha_{AB}(r_{A/B})_y - \omega_{AB}^2\,(r_{A/B})_x \quad \text{and}$$
$$(a_A)_y = (a_B)_y + \alpha_{AB}(r_{A/B})_x - \omega_{AB}^2\,(r_{A/B})_y.$$

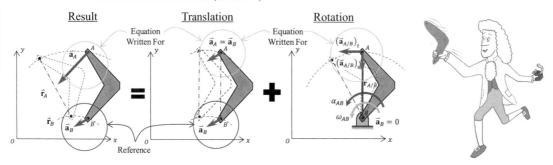

Figure 16.2: Acceleration diagram of Newtdog's boomerang (© E. Diehl).

16.3   ACCELERATION DIAGRAM

Just as we did for Velocity Analysis of rigid bodies, we use a three-part diagram for Acceleration Analysis as shown in Figure 16.2 for the boomerang Newtdog has been throwing. Note that for this image to be accurate the boomerang should be analyzed while it is still in his hand, otherwise it would not have translational or tangential acceleration (but would still have normal acceleration) without applied forces and moments.

To apply the two-dimensional rigid body acceleration equation ($\vec{\mathbf{a}}_A = \vec{\mathbf{a}}_B + \alpha\hat{\mathbf{k}} \times \vec{\mathbf{r}}_{A/B} - \omega^2\vec{\mathbf{r}}_{A/B}$) to the Acceleration Diagram, let's first take a closer look at its parts. If some of these terms appear somewhat familiar, that's because we can compare them to the acceleration equation we used in polar coordinates for particle kinematics: $\vec{\mathbf{a}} = \left(\ddot{r} - r\dot{\theta}^2\right)\hat{\mathbf{e}}_r + \left(r\ddot{\theta} + 2\dot{r}\dot{\theta}\right)\hat{\mathbf{e}}_\theta$. Remember this is a rigid body, so the distance between points doesn't change, thus $\dot{r} = 0$ and $\ddot{r} = 0$. We can see the terms left are quite similar: $\alpha\hat{\mathbf{k}} \times \vec{\mathbf{r}}_{A/B} \to r\ddot{\theta}$ and $-\omega^2\vec{\mathbf{r}}_{A/B} \to -r\dot{\theta}^2$. We refer to these terms as tangential and normal "components," $\left(\vec{\mathbf{a}}_{A/B}\right)_t$ and $\left(\vec{\mathbf{a}}_{A/B}\right)_n$, respectively, although they are vectors not just components. We rewrite the acceleration equation for two dimensions as:

$$\boxed{\vec{\mathbf{a}}_A = \vec{\mathbf{a}}_B + \left(\vec{\mathbf{a}}_{A/B}\right)_t + \left(\vec{\mathbf{a}}_{A/B}\right)_n}$$

where

$$\left(\vec{\mathbf{a}}_{A/B}\right)_t = \alpha_{AB}\hat{\mathbf{k}} \times \vec{\mathbf{r}}_{A/B} \qquad \left|\left(\vec{\mathbf{a}}_{A/B}\right)_t\right| = \left(a_{A/B}\right)_t = \alpha_{AB}r_{A/B}$$

$$\left(\vec{\mathbf{a}}_{A/B}\right)_n = -\omega_{AB}^2\vec{\mathbf{r}}_{A/B} \qquad \left|\left(\vec{\mathbf{a}}_{A/B}\right)_n\right| = \left(a_{A/B}\right)_n = \omega_{AB}^2r_{A/B}.$$

This is the equation we will refer back to most often for rigid body Acceleration Analysis using components and an Acceleration Diagram.

Here is the tricky part of this method. Even though we've designated the relative accelerations as tangential and normal, which we typically associate with components, they are still

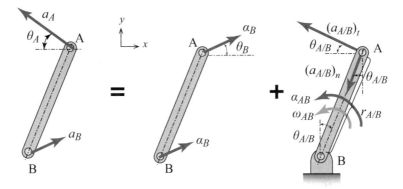

Figure 16.3: Generic acceleration diagram of a link.

vectors. So we need to break them down into their components… we might even say "the components of the components" to remember these must be broken apart. The x-direction (x-dir) and y-direction (y-dir) can be designated as:

x-dir:

$$(a_A)_x = (a_B)_x + \left((a_{A/B})_t\right)_x + \left((a_{A/B})_n\right)_x$$

y-dir:

$$(a_A)_y = (a_B)_y + \left((a_{A/B})_t\right)_y + \left((a_{A/B})_n\right)_y.$$

We need to be careful about signs and angles when applying this method. Unlike the vector math method, we use the diagram rather than a pre-determined convention. While this can be a source for mistakes, it has the advantage of allowing us to visualize what is actually occurring and the relative contribution of each part to the results.

A generic problem is shown in Figure 16.3 and solved using the Vector Components of the Acceleration Diagram method as follows:

$$\vec{a}_A = \vec{a}_B + \left(\vec{a}_{A/B}\right)_t + \left(\vec{a}_{A/B}\right)_n.$$

x-dir:

$$(a_A)_x = (a_B)_x + \left((a_{A/B})_t\right)_x + \left((a_{A/B})_n\right)_x$$

$$- a_A \cos\theta_A = a_B \cos\theta_B - (a_{A/B})_t \cos\theta_{A/B} - (a_{A/B})_n \sin\theta_{A/B}$$

$$- a_A \cos\theta_A = a_B \cos\theta_B - \alpha_{AB}r_{A/B} \cos\theta_{A/B} - \omega_{AB}^2 r_{A/B} \sin\theta_{A/B}.$$

y-dir:

$$(a_A)_y = (a_B)_y + \left((a_{A/B})_t\right)_y + \left((a_{A/B})_n\right)_y$$

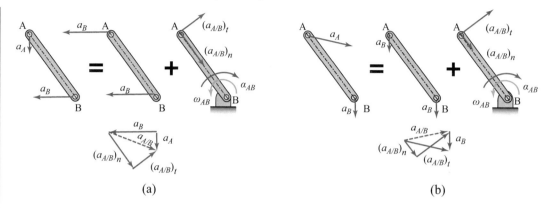

Figure 16.4: Acceleration diagram scenarios where (a) vector diagram forms a quadrilateral and (b) vector diagram forms a bow-tie.

$$a_A \sin \theta_A = a_B \sin \theta_B + \left(a_{A/B}\right)_t \sin \theta_{A/B} - \alpha_{AB} r_{A/B} \cos \theta_{A/B}$$

$$a_A \sin \theta_A = a_B \sin \theta_B + \alpha_{AB} r_{A/B} \sin \theta_{A/B} - \omega_{AB}^2 r_{A/B} \cos \theta_{A/B}.$$

Note the signs are determined by the direction of the vectors drawn on the diagram. We often don't know the direction of the angular acceleration (and therefore don't know the direction of the tangential part of the relative acceleration), but in many cases we will be able to either apply logic to decide which direction it must go to satisfy what is known about the other vectors or we can assume a direction and change our minds if we get a negative result. We will point this out during some examples.

16.4   USING VECTOR TRIANGLES FOR ACCELERATION ANALYSIS

Accelerations *can* be analyzed with vectors triangles, however this process can often lead to confusion. If we use the relative motion equation, $\vec{a}_B = \vec{a}_A + \vec{a}_{B/A}$, we get a triangle, but the relative motion vector is made up of two vectors itself, tangential and normal, so we really have four vectors and these don't necessarily form a quadrilateral. Figure 16.4 shows two similar links with different motion. Below each is the corresponding acceleration vector diagram.

The most obvious issue in Figure 16.4 is that (a) forms a quadrilateral while (b) forms a "bow-tie." The dashed lines in the vector diagrams are the relative acceleration. In both cases we can see two sets of triangles that share the relative acceleration vector as a side. Because of this potential confusion, we'll avoid using vector triangles to solve problems. The triangles remain useful to check results.

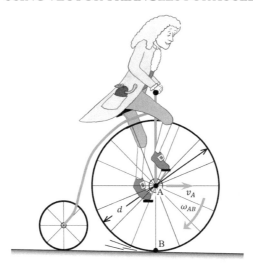

Figure 16.5: Newtdog on a penny-farthing (repeat of Figures 14.10 and 15.8) (© E. Diehl).

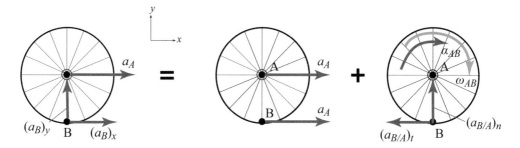

Figure 16.6: Acceleration diagram of Example 16.1.

Another thing of note in Figure 16.4 is the angular velocity and angular acceleration aren't in the same direction which happens when the link is decelerating.

Example 16.1   (repeat of Examples 14.1 and 15.1)

Newtdog is riding a penny-farthing bicycle in Figure 16.5 which has a $d = 4$ ft front wheel. The bicycle is traveling at $v_A = 4$ ft/s, and the front wheel, which is slipping, is rotating at $\omega_{AB} = 3$ rad/s clockwise. The bicycle is also accelerating at $a_A = 1$ ft/s^2 while Newtdog increases his pedaling rate by $\alpha_{AB} = 3$ rad/s^2. Find the total acceleration of Point B on the wheel.

We create the acceleration diagram of the wheel in Figure 16.6 and write the equation for Point B, using Point A as the reference:

$$\vec{a}_B = \vec{a}_A + \left(\vec{a}_{B/A}\right)_t + \left(\vec{a}_{B/A}\right)_n.$$

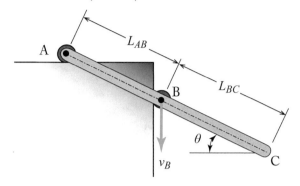

Figure 16.7: Example 16.2 (repeat of Figures 14.12 and 15.10).

x-dir:

$$(a_B)_x = (a_A)_x + \left((a_{B/A})_t\right)_x + \left((a_{B/A})_n\right)_x$$

$$(a_B)_x = a_A - \alpha_{AB}r_{B/A} + (0) = (1) - (3)\,(2) + (0) = -5.000 \text{ ft/s}^2.$$

y-dir:

$$(a_B)_y = (a_A)_y + \left((a_{B/A})_t\right)_y + \left((a_{B/A})_n\right)_y$$

$$(a_B)_y = (0) + (0) + \omega_{AB}^2 r_{B/A} = (0) + (0) + (3)^2\,(2) = 18.00 \text{ ft/s}^2$$

$$a_B = \sqrt{(a_B)_x^2 + (a_B)_y^2} = \sqrt{(-5.000)^2 + (-18.00)^2} = 18.68 \text{ ft/s}^2$$

$$\tan^{-1}\left(\frac{18.00}{5.000}\right) = 74.48°.$$

Answer: $\boxed{\overrightarrow{\mathbf{a}}_B = 18.7 \text{ ft/s}^2 \searrow 74.5°}$.

Example 16.2   (repeat of Examples 14.2 and 15.2)
Link ABC shown in Figure 16.7 has wheels at Points A and B that remain in contact on the horizontal and vertical surfaces. The dimensions are $L_{AB} = 4$ ft and $L_{BC} = 6$ ft. When $\theta = 25°$, the velocity of wheel B is $v_B = 1.5$ ft/s downward but decelerating at $a_B = 0.5$ ft/s^2. Using the velocity results found in Examples 14.2 and 15.2, determine the total acceleration of Point C.

In Examples 14.2 and 15.2 we found the angular velocity was:

$$\omega_{ABC} = 0.4138 \text{ rad/s} \circlearrowleft .$$

If we hadn't already found this in a previous example we would need to do a velocity analysis first to get it. We will solve this problem using Vector Math and the Three Diagram Methods.

Vector Math:

$$\vec{a}_A = (a_A)_x \hat{i} + (0)\,\hat{j} = (a_A)_x \hat{i} + (0)\,\hat{j}$$

$$\vec{a}_B = (0)\hat{i} + (a_B)_y\,\hat{j} = (0)\hat{i} + (0.5)\,\hat{j}$$

$$\vec{\omega}_{AB} = -(0.4138)\,\hat{k}\ \text{rad/s}$$

$$\vec{r}_{A/B} = -(L_{AB}\cos\theta)\,\hat{i} + (L_{AB}\sin\theta)\,\hat{j}$$
$$= -\left((4)\cos\left(25°\right)\right)\hat{i} + \left((4)\sin\left(25°\right)\right)\,\hat{j} = -(3.625)\hat{i} + (1.690)\,\hat{j}.$$

Applied to:

$$\vec{a}_A = \vec{a}_B + \alpha\hat{k}\times\vec{r}_{A/B} - \omega^2\vec{r}_{A/B}.$$

We get:

$$(a_A)_x\hat{i} + (0)\,\hat{j} = (0)\hat{i} + (0.5)\,\hat{j} + \alpha_{AB}\hat{k}\times\left(-(3.625)\hat{i} + (1.690)\,\hat{j}\right)$$
$$- (-(0.4138))^2\left(-(3.625)\hat{i} + (1.690)\,\hat{j}\right)$$

$$(a_A)_x\hat{i} + (0)\,\hat{j} = (0)\hat{i} + (0.5)\,\hat{j} + \alpha_{AB}\,(-(3.625))\,\underbrace{\hat{k}\times\hat{i}}_{+\hat{j}} + \alpha_{AB}\,(1.690)\,\underbrace{\hat{k}\times\hat{j}}_{-\hat{i}}$$
$$- (-(0.4138))^2\,(-(3.625))\,\hat{i} - (-(0.4138))^2\,(1.690)\,\hat{j}.$$

We write the $\hat{j}$ component first to find the angular acceleration:

$$\hat{j}:\ (0) = (0.5) + \alpha_{AB}\,(-(3.625)) - (-(0.4138))^2(1.690)$$

$$\alpha_{AB} = 0.05808\ \text{rad/s}^2.$$

Note this is positive and therefore counter-clockwise.
We don't need to know the $\hat{i}$ component but find it anyway:

$$\hat{i}:\ (a_A)_x = (0) - (0.05808)\,(1.690) - (-(0.4138))^2\,(-(3.625)) = 0.5226\ \text{ft/s}^2.$$

For Point C:

$$\vec{r}_{C/B} = (L_{BC}\cos\theta)\,\hat{i} - (L_{BC}\sin\theta)\,\hat{j}$$
$$= \left((6)\cos\left(25°\right)\right)\hat{i} - \left((6)\sin\left(25°\right)\right)\hat{j} = (5.438)\hat{i} - (2.536)\hat{j}\ \text{ft}.$$

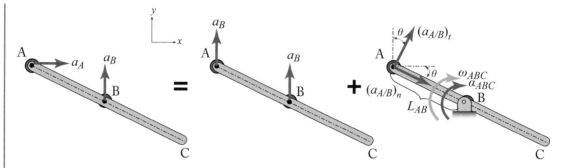

Figure 16.8: Acceleration diagram for Example 16.2 to find angular acceleration.

Applied to:

$$\vec{a}_C = \vec{a}_B + \alpha \hat{k} \times \vec{r}_{C/B} - \omega^2 \vec{r}_{C/B}.$$

We get:

$$(a_C)_x \hat{i} + (a_C)_y \hat{j} = (0)\hat{i} + (0.5)\hat{j} + (0.05808)\hat{k} \times \left((5.438)\hat{i} - (2.536)\hat{j}\right)$$
$$- (-(0.4138))^2 \left((5.438)\hat{i} - (2.536)\hat{j}\right)$$

$$(a_C)_x \hat{i} + (a_C)_y \hat{j} = (0)\hat{i} + (0.5)\hat{j} + (0.05808)(5.438)\underbrace{\hat{k} \times \hat{i}}_{+\hat{j}} - (0.05808)(2.536)\underbrace{\hat{k} \times \hat{j}}_{-\hat{i}}$$
$$- (-(0.4138))^2 ((5.438))\hat{i} - (-(0.4138))^2 (-(2.536))\hat{j}$$

$$\hat{i}: \quad (a_C)_x = (0) + (0.05808)(2.536) - (-(0.4138))^2 ((5.438)) = -0.7839 \text{ ft/s}^2$$
$$\hat{j}: \quad (a_C)_y = (0.5) + (0.05808)(5.438) - (-(0.4138))^2 (-(2.536)) = 1.250 \text{ ft/s}^2$$
$$\vec{a}_C = (-0.7839)\hat{i} + (1.250)\hat{j} \text{ ft/s}^2.$$

Three Diagram Method:
We create the acceleration diagram in Figure 16.8 to first find the angular acceleration:

$$\vec{a}_A = \vec{a}_B + \left(\vec{a}_{A/B}\right)_t + \left(\vec{a}_{A/B}\right)_n.$$

A useful strategy is to first solve the direction whose component is zero in order to find the angular acceleration. Note that the diagram assumes clockwise angular acceleration. While that would be negative in terms of vectors, we use the diagram to assign positive and negative signs, so the tangential component is positive in both directions until we determine if the assumption is correct.

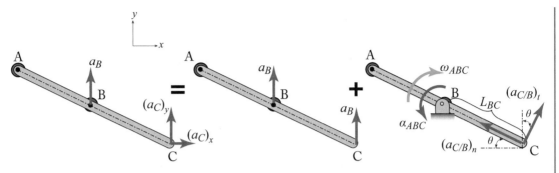

Figure 16.9: Acceleration diagram for Example 16.2 to find acceleration of Point C.

y-dir:

$$(a_A)_y = (a_B)_y + ((a_{A/B})_t)_y + ((a_{A/B})_n)_y$$

$$(0) = (a_B)_y + \alpha_{ABC} L_{AB} \cos\theta - \omega_{ABC}^2 L_{AB} \sin\theta$$

$$(0) = (0.5) + \alpha_{ABC}(4)\cos(25°) - (0.4138)^2(4)\sin(25°)$$

$$\alpha_{ABC} = -0.05808 \text{ rad/s}^2.$$

The negative answer shows we assumed the direction of angular acceleration incorrectly (it rotates counter-clockwise). This agrees with the vector math results. We'll use the negative result along with the diagram as drawn to determine the acceleration of Point A, even though we weren't asked to find it.

x-dir:

$$(a_A)_x = (a_B)_x + ((a_{A/B})_t)_x + ((a_{A/B})_n)_x$$

$$(a_A)_x = a_A = (0) - \alpha_{ABC} L_{AB} \sin\theta + \omega_{ABC}^2 L_{AB} \cos\theta$$

$$a_A = (0) + (-0.05808)(4)\sin(25°) + (0.4138)^2(4)\cos(25°) = 0.5226 \text{ ft/s}^2.$$

This agrees with vector math results. To find the acceleration of Point C, we draw a separate acceleration diagram in Figure 16.9. We could have included this information on the previous acceleration diagram, but that might become somewhat confusing, especially since we already acknowledged we'd guessed the angular acceleration incorrectly. It's often better to use more space to solve problems such as drawing an extra diagram, rather than trying to save paper by drawing a cluttered sketch. We write out the equation for Point C using Point B as the reference:

$$\vec{a}_C = \vec{a}_B + (\vec{a}_{C/B})_t + (\vec{a}_{C/B})_n.$$

In the diagram and the equations we assume the components of Point C are both positive, recognizing negative results will tell use the actual direction.

x-dir:

$$(a_C)_x = (a_B)_x + ((a_{C/B})_t)_x + ((a_{C/B})_n)_x$$

$$(a_C)_x = (0) + \alpha_{ABC} L_{BC} \sin\theta - \omega_{ABC}^2 L_{BC} \cos\theta$$

$$(a_C)_x = (0) + (0.05808)(6)\sin(25°) - (0.4138)^2(6)\cos(25°) = -0.7839 \text{ ft/s}^2.$$

y-dir:

$$(a_C)_y = (a_B)_y + ((a_{C/B})_t)_y + ((a_{C/B})_n)_y$$

$$(a_C)_y = (a_B)_y + \alpha_{ABC} L_{BC} \cos\theta + \omega_{ABC}^2 L_{BC} \sin\theta$$

$$(a_C)_y = (0.5) + (0.05808)(6)\cos(25°) + (0.4138)^2(6)\sin(25°) = 1.250 \text{ ft/s}^2$$

$$\vec{a}_C = (-0.7839)\,\hat{i} + (1.250)\,\hat{j} \text{ ft/s}^2.$$

These agree with the answers from vector math:

$$a_C = \sqrt{(a_C)_x^2 + (a_C)_y^2} = \sqrt{(-0.7839)^2 + (1.250)^2} = 1.476 \text{ ft/s}^2$$

$$\tan^{-1}\left(\frac{1.250}{0.7839}\right) = 57.91°.$$

Answer: $\boxed{\vec{a}_C = 1.48 \text{ ft/s}^2 \searrow 57.9°}$.

We'll use the acceleration diagram method for the remainder of example solutions.

Example 16.3   (repeat of Examples 14.3 and 15.3)

Crank BC of the slider-crank in Figure 16.10 rotates at a constant $\omega_{BC} = 8$ rad/s clockwise $\theta = 40°$. The dimensions are $L_{AB} = 500$ mm, $L_{BC} = 300$ mm, and $L_{BD} = 150$ mm. Determine the acceleration of Point D.

We found $\beta = 22.69°$ in Example 14.3. $\omega_{ABD} = 3.985$ rad/s $\circlearrowleft$.

If we hadn't already found this in a previous example we would need to do a velocity analysis first to get it.

We start by finding the acceleration of Point B (Figure 16.11) which is entirely due to normal acceleration because the crank (Link BC) is rotating at a constant angular speed:

$$a_B = a_{B/C} = \omega_{BC}^2 L_{BC} = (8)^2(0.300) = 1.920 \text{ m/s}^2.$$

We use this known acceleration and apply it to the connecting rod (Link AB) in order to find the angular acceleration. We first draw an acceleration diagram of the connecting rod in Figure 16.12 to focus on Points A and B with the strategy that since Point A only moves in the horizontal direction we're able to find the angular acceleration (which we assume is counter-clockwise).

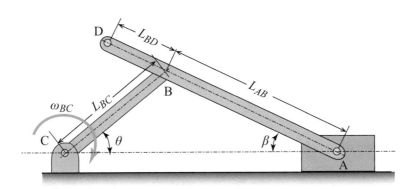

Figure 16.10: Example 16.3 (repeat of Figures 14.16 and 15.12).

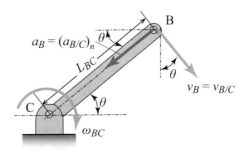

Figure 16.11: Acceleration diagram of Example 16.3 crank.

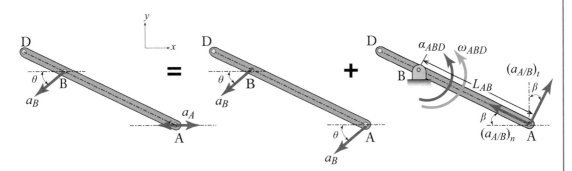

Figure 16.12: Acceleration diagram of Example 16.3 connecting rod to find angular acceleration.

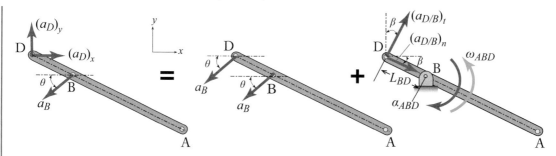

Figure 16.13: Acceleration diagram of Example 16.3 of connecting rod to find acceleration of Point D.

$$\vec{a}_A = \vec{a}_B + \left(\vec{a}_{A/B}\right)_t + \left(\vec{a}_{A/B}\right)_n.$$

y-dir:

$$(a_A)_y = (a_B)_y + \left(\left(a_{A/B}\right)_t\right)_y + \left(\left(a_{A/B}\right)_n\right)_y$$

$$(0) = -a_B \sin\theta + \alpha_{ABD} L_{AB} \cos\beta + \omega_{ABD}^2 L_{AB} \sin\beta$$

$$(0) = -(1.920) \sin\left(40°\right) + \alpha_{ABC} (0.500) \cos\left(22.69°\right) + (3.985)^2 (0.500) \sin\left(22.69°\right)$$

$$\alpha_{ABD} = -20.11 \text{ rad/s}^2.$$

The negative result indicates we've guessed the angular acceleration incorrectly. We'll find the acceleraiton of Point A even thought it isn't asked.

x-dir:

$$(a_A)_x = (a_B)_x + \left(\left(a_{A/B}\right)_t\right)_x + \left(\left(a_{A/B}\right)_n\right)_x$$

$$(a_A)_x = a_A = -a_B \cos\theta + \alpha_{ABC} L_{AB} \sin\beta - \omega_{ABD}^2 L_{AB} \cos\beta$$

$$a_A = -(1.920) \cos\left(40°\right) + (-20.11)(0.500) \sin\left(22.69°\right) - (3.985)^2 (0.500) \cos\left(22.69°\right)$$
$$= -12.68 \text{ m/s}^2.$$

So the piston is decreasing in speed as it approaches the right. This makes sense since it will stop/pause once it reaches the furthest position to the right (called "Top Dead Center" in terms of engines).

To find the acceleration of Point D, we draw a new acceleration diagram in Figure 16.13, this time with the known angular acceleration direction:

$$\vec{a}_D = \vec{a}_B + \left(\vec{a}_{D/B}\right)_t + \left(\vec{a}_{D/B}\right)_n.$$

x-dir:

$$(a_D)_x = (a_B)_x + ((a_{D/B})_t)_x + ((a_{D/B})_n)_x$$

$$(a_D)_x = -a_B \cos\theta + \alpha_{ABD} L_{BD} \sin\beta + \omega_{ABD}^2 L_{BD} \cos\beta$$

$$(a_D)_x = -(1.920)\cos(40°) + (20.11)(0.150)\sin(22.69°)$$
$$+ (3.985)^2 (0.150)\cos(22.69°) = 1.890 \text{ m/s}^2.$$

y-dir:

$$(a_D)_y = (a_B)_y + ((a_{D/B})_t)_y + ((a_{D/B})_n)_y$$

$$(a_D)_y = -a_B \sin\theta + \alpha_{ABD} L_{BD} \cos\beta - \omega_{ABD}^2 L_{BD} \sin\beta$$

$$(a_D)_y = -(1.920)\sin(40°) + (20.11)(0.150)\cos(22.69°)$$
$$- (3.985)^2 (0.150)\sin(22.69°) = 0.6300 \text{ m/s}^2$$

$$\vec{a}_D = (1.890)\,\hat{i} + (0.6300)\,\hat{j} \text{ m/s}^2$$

$$a_D = \sqrt{(a_D)_x^2 + (a_D)_y^2} = \sqrt{(1.890)^2 + (0.6300)^2} = 1.993 \text{ m/s}^2$$

$$\tan^{-1}\left(\frac{0.6300}{1.890}\right) = 18.43°.$$

Answer: $\boxed{\vec{a}_D = 1.99 \text{ m/s}^2 \nearrow 18.4°}$.

Example 16.4 (repeat of Examples 14.4 and 15.4)

In the position shown, bar CD of the four-bar mechanism in Figure 16.14 has an angular velocity of $\omega_{CD} = 12$ rad/s counter-clockwise and accelerating at $\alpha_{CD} = 25$ rad/s^2 counter-clockwise. The dimensions are $L_{AB} = 600$ mm, $L_{BC} = 400$ mm, $L_{CD} = 350$ mm, and $x_{AD} = 200$ mm. Determine the angular acceleration of Rocker AB.

From Examples 14.4 and 15.4 we found the angle of Rocker AB to be $\theta = 23.56°$ and the angular velocities of the rocker (Link AB) and coupler (Link BC) to be:

$$\omega_{AB} = 7.637 \text{ rad/s} \circlearrowleft \quad \text{and} \quad \omega_{BC} = 4.579 \text{ rad/s} \circlearrowright .$$

The acceleration of Point C of the Crank CG is found from the given angular velocity and acceleration in pure rotation using Figure 16.15a:

$$(a_C)_x = (a_{C/D})_n = \omega_{CD}^2 L_{CD} = (12)^2 (0.350) = 50.40 \text{ m/s}^2 \leftarrow$$

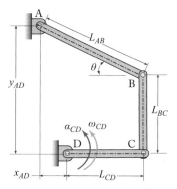

Figure 16.14: **Example** 16.4 (repeat of Figures 14.20 and 15.14).

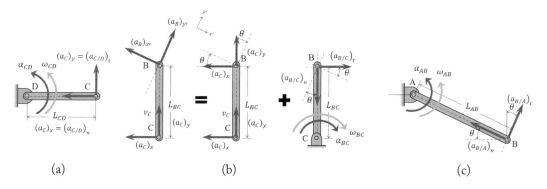

Figure 16.15: Acceleration diagrams of (a) crank CD, (b) rocker AB, and (c) coupler BC of Example 16.4.

$$(a_C)_y = (a_{C/D})_t = \alpha_{CD} L_{CD} = (25)\,(0.350) = 8.750 \text{ m/s}^2 \uparrow .$$

We find the normal and tangential acceleration of Rocker AB using the acceleration diagram of Figure 16.15c, leaving the desired angular acceleration as a variable.

$$\vec{a}_B = \vec{a}_{B/A} = (\vec{a}_{B/A})_t + (\vec{a}_{B/A})_n,$$

where

$$(a_{B/A})_n = \omega_{AB}^2 L_{AB} = (7.637)^2\,(0.600) = 34.99 \text{ m/s}^2 \searrow 23.56°$$

$$(a_{B/A})_t = \alpha_{AB} L_{AB} = \alpha_{AB}\,(0.600) \,|\nearrow\ 23.56° \quad \textcircled{1}.$$

For link BC using the acceleration diagram of Figure 16.15b, we have:

$$\vec{a}_B = \vec{a}_C + (\vec{a}_{B/C})_t + (\vec{a}_{B/C})_n.$$

We apply a bit of strategy here; since coupler BC is vertical, we can ignore its angular acceleration if we stay in the y-direction only.

y-dir:

$$((a_{B/A})_t)_y + ((a_{B/A})_n)_y = (a_C)_y + ((a_{B/C})_t)_y + ((a_{B/C})_n)_y$$

$$(\alpha_{AB}L_{AB})_t \cos\theta + \omega_{AB}^2 L_{AB} \sin(\theta) = (a_C)_y + (0) - \omega_{BC}^2 L_{BC}$$

$$\alpha_{AB}(0.600)\cos(23.56°) + (34.99)\sin(23.56°) = (8.750) + (0) - (4.579)^2(0.400)$$

$$\alpha_{AB} = -24.77 \text{ rad/s}^2 = 24.77 \text{ rad/s}^2 \circlearrowright .$$

Note that since we obtained a negative value, the assumed direction of angular acceleration of link AB was incorrect and is clockwise. While this answers the question, we'll also need the angular acceleration of link BC for an example in Class 21. We use the acceleration diagram of Figure 16.5c in the x-direction even though we now know that the acceleration $(a_{B/A})_t$ is in the opposite direction and use the negative value for α_{AB}.

x-dir:

$$((a_{B/A})_t)_x + ((a_{B/A})_n)_x = (a_C)_x + ((a_{B/C})_t)_x + ((a_{B/C})_n)_x$$

$$\alpha_{AB}L_{AB}\sin\theta - \omega_{AB}^2 L_{AB}\cos\theta = -(a_C)_x + \alpha_{BC}L_{BC} + (0)$$

$$(-24.77)(0.600)\sin(23.56°) - (34.99)\cos(23.56°) = -(50.40) + \alpha_{BC}(0.400) + (0)$$

$$\alpha_{BC} = 30.99 \text{ rad/s}^2 \circlearrowright .$$

Answer: $\boxed{\vec{\alpha}_{AB} = 24.8 \text{ rad/s}^2 \circlearrowright}$.

Rigid body acceleration problems are complicated to solve having multiple steps that require focus and care to avoid mistakes. Quite often problems assigned for homework or exams will have some simplifying features (such as coupler BC being vertical, allowing us to ignore its tangential component in the previous example) that will appear to be "tricks" in the sense that they are obvious once you see the solution. Students often struggle with these types of problems, so a second chapter has been included to provide more examples (as well as introduce some alternate approaches). It is strongly encouraged that you use the problems in this Class as practice. Write down the problem and cover up the solution, don't peek, and try to solve it yourself to completion. Doing so will build up your problem-solving muscles in the most effective way. Only through experience can you gain insight into the solution strategies.

Book 2 - Class 17

https://www.youtube.com/watch?v=OLzRH6AvPu8

C L A S S   17

Acceleration Analysis (Part 2)

B.L.U.F. (Bottom Line Up Front)

- Another way to write vectors is with complex number notation.

- Vector Loops can be used to find the position, velocity, and acceleration of linkages using complex number notation.

17.1 MORE ACCELERATION ANALYSIS OF RIGID BODIES

This class includes an additional example of rigid body acceleration analysis applied to a well-established mechanism. It's intended to be more challenging than the examples in Class 16 and provide us with an opportunity to practice on another real world application.

Example 17.1
Figure 17.1 shows a particular four-bar configuration called a Hoeken straight-line mechanism. A point on an extension of the coupler produces a straight line for a portion of the full rotation path. This property can be useful in certain machines including one that pushes something across a surface and then lifts up on its return. The dimensions are $L_{AB} = 1$ in, $L_{BC} = 2.5$ in, $L_{CD} = 2.5$ in, $L_{CE} = 2.5$ in, and $L_{AE} = 2$ in (these proportions determine this motion). Crank AB rotates counter-clockwise at $\dot{\theta}_{AB} = 10$ rad/s and $\ddot{\theta}_{AB} = 2$ rad/s^2. Determine the velocity and acceleration of Point D when crank AB is horizontal as shown. Note that we expect the velocity to be completely horizontal since this is the straight-line portion.

Position Analysis:

The first step is to perform a position analysis to find the angles of links BC and CE. Figure 17.2 shows a breakdown of the dimensions of the links:

To find the unknown angles we apply the Law of Cosines ($c^2 = a^2 + b^2 - 2ab \cos C$) because we know the lengths of the sides of the triangle formed by the links:

$$L_{CE}^2 = L_{BC}^2 + (L_{AB} + L_{AE})^2 - 2L_{BC}(L_{AB} + L_{AE}) \cos \theta_{BC}$$

$$\theta_{BC} = \cos^{-1}\left[\frac{L_{BC}^2 + (L_{AB} + L_{AE})^2 - L_{CE}^2}{2L_{BC}(L_{AB} + L_{AE})}\right]$$

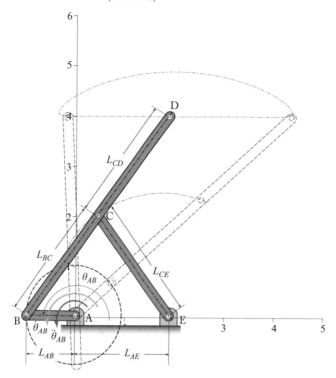

Figure 17.1: Hoeken straight-line mechanism of Example 17.1.

$$\theta_{BC} = \cos^{-1}\left[\frac{(2.5)^2 + ((1) + (2))^2 - (2.5)^2}{2\,(2.5)\,((1) + (2))}\right] = 53.13° \; .$$

Because the sides form an equilateral triangle, we know $\beta = \theta_{BC} = 53.13°$. The third angle is $\vartheta = 180° - \beta - \theta_{BC} = 180° - 2\,(53.13°) = 73.74°$.

We note that in this position Point D is directly above Point E. The distance L_{ED} is: $L_{ED} = (L_{BC} + L_{CD})\sin\theta_{BC} = ((2.5) + (2.5))\sin(53.13°) = 4.000$ in. This could have been found from inspection as a 3–4–5 triangle.

Velocity Analysis:

The velocity analysis is most easily performed using the Instantaneous Center of Rotation. Figure 17.3 shows the velocity diagrams of all three moving links.

The velocity of Point B:

$$v_B = L_{AB}\dot{\theta}_{AB} = (1)\,(10) = 10 \text{ in/s}.$$

We note that the ICR of link BCD is located at Point E. The ICR on four-bar linkages lands on the intersection of lines drawn through the links which in this case is at Point E.

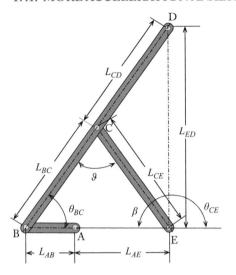

Figure 17.2: Linkage geometry in Example 17.1.

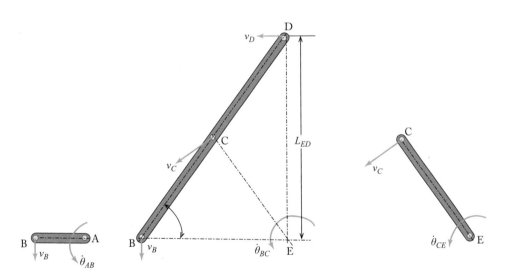

Figure 17.3: Velocity diagrams of Example 17.1.

The angular velocity of link BC is:

$$\dot{\theta}_{BC} = \frac{v_B}{(L_{AB} + L_{AE})} = \frac{(10)}{((1) + (2))} = 3.333 \text{ rad/s.}$$

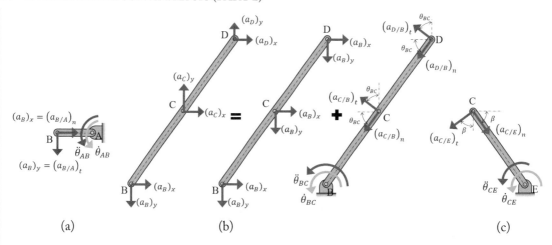

(a)   (b)   (c)

Figure 17.4: Acceleration diagrams of Example 17.1.

We will want to know the angular velocity of link CE as well, and again since the ICR lands on Point E, the angular velocity of link CE is also $\dot{\theta}_{CE} = \dot{\theta}_{BC} = 3.333$ rad/s.

The velocity of Point D is $v_D = L_{ED}\dot{\theta}_{BC} = (4)(3.333) = 13.33$ in/s $\leftarrow$, so it is completely horizontal as expected. Note that in the real Hoeken straight line mechanism there are very slight changes in the direction of the velocity as it passes through the flat portion as this is, in fact, an approximate straight line mechanism. In this crank position, however, the velocity is completely horizontal.

Acceleration Analysis:

Acceleration diagrams for the three moving links are shown in Figure 17.4. We march through the links, beginning with the acceleration of Point B on link AB:

$$(a_B)_x = (a_{B/A})_n = \dot{\theta}_{AB}^2 L_{AB} = (10)^2 (1) = 100 \text{ in/s}^2 \rightarrow$$

$$(a_B)_y = (a_{B/A})_t = \ddot{\theta}_{AB} L_{AB} = (2)(1) = 2 \text{ in/s}^2 \downarrow .$$

The acceleration analysis of Point C on rocker CE has the normal acceleration known and the angular acceleration unknown:

$$(a_{C/E})_n = \dot{\theta}_{CE}^2 L_{CE} = (3.333)^2 (2.5) = 27.77 \text{ in/s}^2 \searrow$$

$$(a_{C/E})_t = \ddot{\theta}_{CE} L_{CE} = \ddot{\theta}_{CE}(2.5).$$

Breaking these into x and y coordinates gives the following two equations in terms of the angular acceleration:

$$(a_C)_x = -(a_{C/E})_t \sin \beta + (a_{C/E})_n \cos \beta$$
$$= -\ddot{\theta}_{CE} (2.5) \sin (53.13°) + (27.77) \cos (53.13°) = -(2.000) \ddot{\theta}_{CE} + (16.67) \quad \textcircled{1}$$

$$(a_C)_y = -(a_{C/E})_t \cos \beta - (a_{C/E})_n \sin \beta$$
$$= -\ddot{\theta}_{CE} (2.5) \cos (53.13°) - (27.77) \sin (53.13°) = -(1.500) \ddot{\theta}_{CE} - (22.22) \quad \textcircled{2}$$

An acceleration analysis of Point C of the coupler BC:

$$\vec{a}_C = \vec{a}_B + (\vec{a}_{C/B})_t + (\vec{a}_{C/B})_n .$$

x-dir:

$$(a_C)_x = (a_B)_x + ((a_{C/B})_t)_x + ((a_{C/B})_n)_x$$
$$(a_C)_x = (a_B)_x - \ddot{\theta}_{BC} L_{BC} \sin \theta_{BC} - \dot{\theta}_{BC}^2 L_{BC} \cos \theta_{BC}$$

$$(a_C)_x = (100) - \ddot{\theta}_{BC} (2.5) \sin (53.13°) - (3.333)^2 (2.5) \cos (53.13°)$$
$$= (83.33) - (2.000) \ddot{\theta}_{BC} \quad \textcircled{3}.$$

y-dir:

$$(a_C)_y = (a_B)_y + ((a_{C/B})_t)_y + ((a_{C/B})_n)_y$$
$$(a_C)_y = -(a_B)_y + \ddot{\theta}_{BC} L_{BC} \cos \theta_{BC} - \dot{\theta}_{BC}^2 L_{BC} \sin \theta_{BC}$$

$$(a_C)_y = -(2) + \ddot{\theta}_{BC} (2.5) \cos (53.13°) - (3.333)^2 (2.5) \sin (53.13°)$$
$$= -(24.22) + (1.500) \ddot{\theta}_{BC} \quad \textcircled{4}.$$

Combine equations $\textcircled{1}$, $\textcircled{2}$, $\textcircled{3}$, and $\textcircled{4}$ and we have two equations and two unknowns:

$$-(2.000) \ddot{\theta}_{CE} + (16.67) = (83.33) - (2.000) \ddot{\theta}_{BC}$$

$$-(1.500) \ddot{\theta}_{CE} - (22.22) = -(24.22) + (1.500) \ddot{\theta}_{BC}$$

$$\ddot{\theta}_{CE} = \frac{-(83.33) + (2.000) \ddot{\theta}_{BC} + (16.67)}{(2.000)} = -(33.33) + \ddot{\theta}_{BC}$$

$$\ddot{\theta}_{CE} = \frac{(24.22) - (2.000) \ddot{\theta}_{BC} - (22.22)}{(1.500)} = (1.333) - \ddot{\theta}_{BC}$$

$$-(33.33) + \ddot{\theta}_{BC} = (1.333) - \ddot{\theta}_{BC}$$

$$\ddot{\theta}_{BC} = 17.33 \text{ rad/s}^2 \circlearrowleft$$

$$\ddot{\theta}_{CE} = -(33.33) + (17.33) = -16.00 \text{ rad/s}^2 = 16.00 \text{ rad/s}^2 \circlearrowleft$$

The acceleration of Point D on the coupler extension:

$$\vec{\mathbf{a}}_D = \vec{\mathbf{a}}_B + \left(\vec{\mathbf{a}}_{D/B}\right)_t + \left(\vec{\mathbf{a}}_{D/B}\right)_n.$$

x-dir:

$$(a_D)_x = (a_B)_x + \left(\left(a_{D/B}\right)_t\right)_x + \left(\left(a_{D/B}\right)_n\right)_x$$

$$(a_D)_x = (a_B)_x - \ddot{\theta}_{BC}\left(L_{BC} + L_{BC}\right)\sin\theta_{BC} - \dot{\theta}_{BC}^2\left(L_{BC} + L_{BC}\right)\cos\theta_{BC}$$

$$(a_D)_x = (100) - (17.33)(5)\sin\left(53.13°\right) - (3.333)^2(5)\cos\left(53.13°\right) = -2.667 \text{ in/s}^2.$$

y-dir:

$$(a_D)_y = (a_B)_y + \left(\left(a_{D/B}\right)_t\right)_y + \left(\left(a_{D/B}\right)_n\right)_y$$

$$(a_D)_y = -(a_B)_y + \ddot{\theta}_{BC}\left(L_{BC} + L_{BC}\right)\cos\theta_{BC} - \dot{\theta}_{BC}^2\left(L_{BC} + L_{BC}\right)\sin\theta_{BC}$$

$$(a_D)_y = -(2) + (17.33)(5)\cos\left(53.13°\right) - (3.333)^2(5)\sin\left(53.13°\right) = 5.556 \text{ in/s}^2.$$

Answer: $\boxed{\vec{\mathbf{a}}_D = (-2.67)\,\hat{\mathbf{i}} + (5.56)\,\hat{\mathbf{j}} \text{ in/s}^2}$.

It is somewhat surprising that the acceleration has a y-component. This is because the "straight line mechanism" is actually an approximation of a straight line. While in this straight line region there are relatively small velocities in the y-direction (although zero in this position) so even a minor change in the velocity is represented by an acceleration.

17.2 POLAR COMPLEX NOTATION

An alternate way to represent vectors is with polar complex notation that is quite useful when applied to the dynamics of linkages. Figure 17.5 shows a typical vector in both Polar Complex Number notation and Cartesian notation. The coefficient represents the magnitude, and the angle, measured counter-clockwise from the right side horizontal, is multiplied by the imaginary number within the exponential. For example, a vector $\vec{A}$ with magnitude a and angle θ_A is written:

$$\vec{A} = a \cdot e^{i\theta_A}.$$

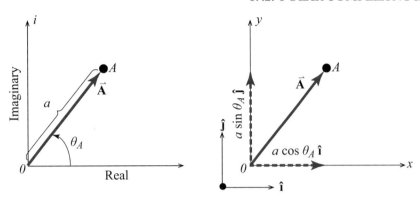

Figure 17.5: Polar complex notation and Cartesian vector representations.

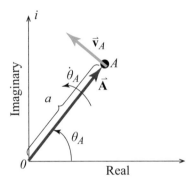

Figure 17.6: Velocity of constant magnitude position vector.

To convert this vector to rectangular components we use the Euler Equivalents Equation ($e^{\pm i\theta} = \cos\theta \pm i\sin\theta$) where the imaginary part of the complex number is the vertical axis. For the example vector in Figure 17.5 this conversion looks like this:

$$\vec{A} = a \cdot e^{i\theta_A} = a\cos\theta_A + ai\sin\theta_A = (a\cos\theta_A)\,\hat{\mathbf{i}} + (a\sin\theta_A)\,\hat{\mathbf{j}}.$$

When a position vector's magnitude remains constant (as it is in a rigid link), we take the time derivative to get velocity

$$\vec{v}_A = \frac{d\vec{A}}{dt} = aie^{i\theta_A}\frac{d\theta_A}{dt} = a\dot{\theta}_A i e^{i\theta_A}.$$

Note that "i" essentially *rotates the velocity 90° from the position vector* (real to imaginary). Figure 17.6 shows the resulting velocity vector on a rigid link position vector which has the magnitude $a\dot{\theta}_A$.

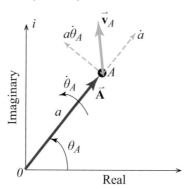

Figure 17.7: Velocity of varying magnitude position vector.

If we want to convert this back into Cartesian coordinates we substitute the Euler Equivalent Equation and assign the $\hat{\mathbf{i}}$ unit vector for the real part and the $\hat{\mathbf{j}}$ unit vector for the imaginary part:

$$\vec{\mathbf{v}}_A = a\dot{\theta}_A i \left(\cos\theta_A + i\sin\theta_A\right) = a\dot{\theta}_A i \cos\theta_A - a\dot{\theta}_A \sin\theta_A$$
$$= \left(-a\dot{\theta}_A \sin\theta_A\right)\hat{\mathbf{i}} + \left(a\dot{\theta}_A \cos\theta_A\right)\hat{\mathbf{j}}.$$

The results make physical sense, since the x-direction is negative as shown in Figure 17.6.

If the position vector changes length (as in Figure 17.7), then the time derivative results in:

$$\vec{\mathbf{v}}_A = \frac{d\vec{A}}{dt} = ae^{i\theta_A}\frac{da}{dt} + aie^{i\theta_A}\frac{d\theta_A}{dt}$$
$$\vec{\mathbf{v}}_A = \dot{a}e^{i\theta_A} + a\dot{\theta}_A i e^{i\theta_A}.$$

We again substitute the Euler Equivalent Equation to get:

$$\vec{\mathbf{v}}_A = \dot{a}\left(\cos\theta_A + i\sin\theta_A\right) + a\dot{\theta}_A i \left(\cos\theta_A + i\sin\theta_A\right)$$
$$= \dot{a}\cos\theta_A + \dot{a}i\sin\theta_A + a\dot{\theta}_A i \cos\theta_A - a\dot{\theta}_A \sin\theta_A$$
$$= \left(\dot{a}\cos\theta_A - a\dot{\theta}_A \sin\theta_A\right)\hat{\mathbf{i}} + \left(\dot{a}\sin\theta_A + a\dot{\theta}_A \cos\theta_A\right)\hat{\mathbf{j}}.$$

If we take another time derivative of the varying magnitude position vector velocity to get acceleration, we see the same terms we've found in particle acceleration analysis:

$$\vec{\mathbf{a}}_A = \frac{d\vec{\mathbf{v}}_A}{dt} = \ddot{a}e^{i\theta_A} + \dot{a}\dot{\theta}_A i e^{i\theta_A} + \dot{a}\dot{\theta}_A i e^{i\theta_A} + a\ddot{\theta}_A i e^{i\theta_A} + a\dot{\theta}_A^2 i^2 e^{i\theta_A}$$

$$\vec{\mathbf{a}}_A = \ddot{a}e^{i\theta_A} - a\dot{\theta}_A^2 e^{i\theta_A} + a\ddot{\theta}_A i e^{i\theta_A} + 2\dot{a}\dot{\theta}_A i e^{i\theta_A}.$$

Compare this result to particle acceleration in polar coordinates:

$$\vec{a} = \left(\ddot{r} - r\dot{\theta}^2 \right) \hat{e}_r + \left(r\ddot{\theta} + 2\dot{r}\dot{\theta} \right) \hat{e}_\theta.$$

We could also substitute the Euler Equivalent Equation into the acceleration above to switch to Cartesian coordinates.

In the next class we will have links that change length, but in the typical rigid body kinematics problems the terms $\dot{a}$ and $\ddot{a}$ are zero. The velocity and acceleration of a rigid link are therefore:

$$\vec{v}_A = a\dot{\theta}_A i\, e^{i\theta_A}$$

$$\vec{a}_A = -a\dot{\theta}_A^2 e^{i\theta_A} + a\ddot{\theta}_A i\, e^{i\theta_A}.$$

In the next sections we'll introduce the idea of a "vector loop" and apply the above polar complex number notation to the two most common simple linkages: four-bar linkages and slider cranks.

17.3   VECTOR LOOPS

We can use vectors to represent the links of mechanisms. Figure 17.8 demonstrates vectors loops for slider-cranks and four-bar linkages. Looking closely at both vector loops we can see that simple vector addition can be written for each as the following equations:

Slider-crank:
$$\vec{A} - \vec{B} - \vec{C} = 0 \qquad a \cdot e^{i\theta_A} - b \cdot e^{i\theta_B} - c = 0.$$

Four-bar linkage:
$$\vec{A} + \vec{B} - \vec{C} - \vec{D} = 0 \qquad a \cdot e^{i\theta_A} + b \cdot e^{i\theta_B} - c \cdot e^{i\theta_C} - d = 0.$$

Note that the equations follow the vectors in a "loop," starting from vector A and moving clockwise, where positive vectors agree with the direction of the loop and negative vectors oppose the loop. Also notice that c in the slider-crank and d in the four-bar are horizontal so don't have exponentials.

To understand the usefulness of these, we'll demonstrate that with the knowns and unknowns we can find the positions of each.

For the slider-crank we substitute the Euler Equivalents Equation into $a \cdot e^{i\theta_A} - b \cdot e^{i\theta_B} - c = 0$ and get:

$$a \cdot \cos \theta_A + a \cdot i \sin \theta_A - b \cdot \cos \theta_B - b \cdot i \sin \theta_B - c = 0.$$

We can separate out the real and imaginary parts:

$$a \cdot \cos \theta_A - b \cdot \cos \theta_B = c$$

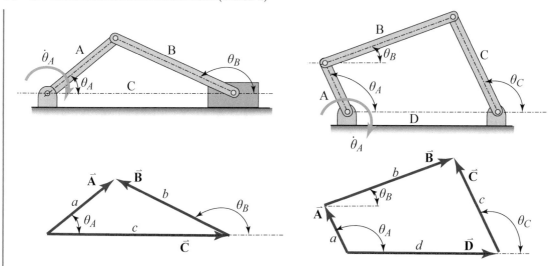

Figure 17.8: Slider-crank and four-bar linkage vector loop equivalent.

$$a \cdot \sin \theta_A - b \cdot \sin \theta_B = 0.$$

The knowns are the lengths a and b, the unknowns are θ_A, θ_B, and c. If we choose to make θ_A an input value we are left with two equations and two unknowns and all that is left is the algebra:

$$\theta_B = \sin^{-1}\left[\frac{a}{b}\sin \theta_A\right]$$

$$c = a \cos \theta_A - b \cdot \cos \theta_B.$$

The position analysis of the four-bar linkage is a bit more challenging. If we substitute the Euler Equivalents Equation into $a \cdot e^{i\theta_A} + b \cdot e^{i\theta_B} - c \cdot e^{i\theta_C} - d = 0$ and we get:

$$a \cdot \cos \theta_A + a \cdot i \sin \theta_A + b \cdot \cos \theta_B + b \cdot i \sin \theta_B - c \cdot \cos \theta_C - c \cdot i \sin \theta_C - d = 0.$$

We can separate out the real and imaginary parts:

$$a \cdot \cos \theta_A + b \cdot \cos \theta_B - c \cdot \cos \theta_C - d = 0$$

$$a \cdot \sin \theta_A + b \cdot \sin \theta_B - c \cdot \sin \theta_C = 0.$$

The knowns are the lengths a, b, c, and d, and the unknowns are θ_A, θ_B, and θ_C. If we choose to make θ_A an input value we are again left with two equations and two unknowns. The algebra for this is a bit more challenging and will not be shown here, but we recognize it is possible to solve. Numerical approaches rather than algebraic are often preferred. There is an additional issue with both a physical and mathematical consequence. For a given input angle θ_A there exist two

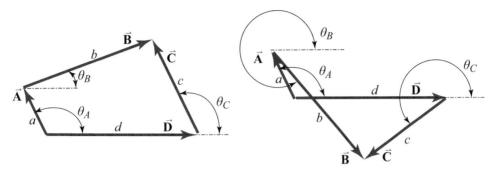

Figure 17.9: Ambiguous position of four-bar linkage with same input angle θ_A.

possible solutions. The bars can be flipped (also called "crossed") so there is an inherent ambiguity as shown in Figure 17.9. Mathematically, this ambiguity arises when taking the inverse sine which can have two results. For instance, the inverse sine of 0.5 can be either 30° or 150°.

We will use these vector loops for rigid body velocity and acceleration in the sections that follow and again in Class 18.

17.4   SLIDER-CRANK VELOCITY AND ACCELERATION ANALYSIS USING VECTOR LOOPS

The slider-crank from Figure 17.8 has a vector loop written as:

$$\vec{A} - \vec{B} - \vec{C} = 0 \qquad a \cdot e^{i\theta_A} - b \cdot e^{i\theta_B} - c = 0$$

$$\frac{d}{dt}\left[\vec{A} - \vec{B} - \vec{C}\right] = 0$$

$$a\dot{\theta}_A i e^{i\theta_A} - b\dot{\theta}_B i e^{i\theta_B} - \dot{c} = 0.$$

Substitute the Euler Equivalents Equation and separate the real and imaginary parts:

$$a\dot{\theta}_A i \left(\cos\theta_A + i\sin\theta_A\right) - b\dot{\theta}_B i \left(\cos\theta_B + i\sin\theta_B\right) - \dot{c} = 0$$

$$a\dot{\theta}_A i \cos\theta_A + a\dot{\theta}_A i\, i \sin\theta_A - b\dot{\theta}_B i \cos\theta_B - b\dot{\theta}_B i\, i \sin\theta_B - \dot{c} = 0$$

$$a\dot{\theta}_A i \cos\theta_A - a\dot{\theta}_A \sin\theta_A - b\dot{\theta}_B i \cos\theta_B + b\dot{\theta}_B \sin\theta_B - \dot{c} = 0.$$

Separating the real and imaginary parts results in two equations:

$$-a\dot{\theta}_A \sin\theta_A + b\dot{\theta}_B \sin\theta_B - \dot{c} = 0$$

$$\dot{\theta}_A \cos\theta_A - b\dot{\theta}_B \cos\theta_B = 0.$$

If the angular velocity of the crank ($\dot{\theta}_A$) is the input, we have two equations and two unknowns ($\dot{\theta}_B$ and $\dot{c}$) and can solve:

$$\dot{\theta}_B = \frac{a \cdot \cos \theta_A}{b \cdot \cos \theta_B} \dot{\theta}_A$$

$$\dot{c} = -a \cdot \dot{\theta}_A \sin \theta_A + b \cdot \dot{\theta}_B \sin \theta_B.$$

The acceleration has more terms to consider. Note that since the links remain the same length, terms like $\ddot{a}$, $\dot{a}$, $\ddot{b}$, and $\dot{b}$ are omitted:

$$\frac{d}{dt} \left[a\dot{\theta}_A i\, e^{i\theta_A} - b\dot{\theta}_B i\, e^{i\theta_B} - \dot{c} \right] = 0$$

$$a\ddot{\theta}_A i\, e^{i\theta_A} + a\dot{\theta}_A^2 i^2 e^{i\theta_A} - b\ddot{\theta}_B i\, e^{i\theta_B} - b\dot{\theta}_B^2 i^2 e^{i\theta_B} - \ddot{c} = 0$$

$$a\ddot{\theta}_A i\, e^{i\theta_A} - a\dot{\theta}_A^2 e^{i\theta_A} - b\ddot{\theta}_B i\, e^{i\theta_B} + b\dot{\theta}_B^2 e^{i\theta_B} - \ddot{c} = 0.$$

If we substitute the Euler Equivalents Equation,

$$a\ddot{\theta}_A i\, (\cos \theta_A + i \sin \theta_A) - a\dot{\theta}_A^2 (\cos \theta_A + i \sin \theta_A) - b\ddot{\theta}_B i\, (\cos \theta_B + i \sin \theta_B)$$
$$+ b\dot{\theta}_B^2 (\cos \theta_B + i \sin \theta_B) - \ddot{c} = 0$$

$$a\ddot{\theta}_A i \cos \theta_A + a\ddot{\theta}_A i\, i \sin \theta_A - a\dot{\theta}_A^2 \cos \theta_A - a\dot{\theta}_A^2 i \sin \theta_A - b\ddot{\theta}_B i \cos \theta_B - b\ddot{\theta}_B i\, i \sin \theta_B$$
$$+ b\dot{\theta}_B^2 \cos \theta_B + b\dot{\theta}_B^2 i \sin \theta_B - \ddot{c} = 0$$

$$a\ddot{\theta}_A i \cos \theta_A - a\ddot{\theta}_A \sin \theta_A - a\dot{\theta}_A^2 \cos \theta_A - a\dot{\theta}_A^2 i \sin \theta_A - b\ddot{\theta}_B i \cos \theta_B + b\ddot{\theta}_B \sin \theta_B$$
$$+ b\dot{\theta}_B^2 \cos \theta_B + b\dot{\theta}_B^2 i \sin \theta_B - \ddot{c} = 0.$$

Separating into the real and imaginary parts results in two equations:

$$- a\ddot{\theta}_A \sin \theta_A - a\dot{\theta}_A^2 \cos \theta_A + b\ddot{\theta}_B \sin \theta_B + b\dot{\theta}_B^2 \cos \theta_B - \ddot{c} = 0$$

$$a\ddot{\theta}_A \cos \theta_A - a\dot{\theta}_A^2 \sin \theta_A - b\ddot{\theta}_B \cos \theta_B + b\dot{\theta}_B^2 \sin \theta_B = 0.$$

If a known input is the crank acceleration, $\ddot{\theta}_A$, we have two equations and two unknowns, $\ddot{\theta}_B$ and $\ddot{c}$:

$$\ddot{\theta}_B = \frac{\left[a\ddot{\theta}_A \cos \theta_A - a\dot{\theta}_A^2 \sin \theta_A + b\dot{\theta}_B^2 \sin \theta_B \right]}{b \cos \theta_B}$$

$$\ddot{c} = -a\ddot{\theta}_A \sin \theta_A - a\dot{\theta}_A^2 \cos \theta_A + b\ddot{\theta}_B \sin \theta_B + b\dot{\theta}_B^2 \cos \theta_B.$$

17.5   FOUR-BAR LINKAGE VELOCITY AND ACCELERATION ANALYSIS USING VECTOR LOOPS

The four-bar linkage from Figure 17.8 has a vector loop written as:

$$\vec{A} + \vec{B} - \vec{C} - \vec{D} = 0 \qquad a \cdot e^{i\theta_A} + b \cdot e^{i\theta_B} - c \cdot e^{i\theta_C} - d = 0$$

$$\frac{d}{dt}\left[\vec{A} + \vec{B} - \vec{C} - \vec{D}\right] = 0.$$

Note that the ground link distance remains constant ($\dot{d} = 0$), so the results are:

$$a\dot{\theta}_A i e^{i\theta_A} + b\dot{\theta}_B i e^{i\theta_B} - c\dot{\theta}_C i e^{i\theta_C} = 0.$$

Substitute the Euler Equivalents Equation and separate the real and imaginary parts:

$$a\dot{\theta}_A i\,(\cos\theta_A + i\sin\theta_A) + b\dot{\theta}_B i\,(\cos\theta_B + i\sin\theta_B) - c\dot{\theta}_C i\,(\cos\theta_C + i\sin\theta_C) = 0$$

$$a\dot{\theta}_A i\cos\theta_A + a\dot{\theta}_A i i\sin\theta_A + b\dot{\theta}_B i\cos\theta_B + b\dot{\theta}_B i i\sin\theta_B - c\dot{\theta}_C i\cos\theta_C - c\dot{\theta}_C i i\sin\theta_C = 0$$

$$a\dot{\theta}_A i\cos\theta_A - a\dot{\theta}_A\sin\theta_A + b\dot{\theta}_B i\cos\theta_B - b\dot{\theta}_B\sin\theta_B - c\dot{\theta}_C i\cos\theta_C + c\dot{\theta}_C\sin\theta_C = 0.$$

Separate the real and imaginary parts:

$$- a\dot{\theta}_A\sin\theta_A - b\dot{\theta}_B\sin\theta_B + c\dot{\theta}_C\sin\theta_C = 0$$

$$a\dot{\theta}_A\cos\theta_A + b\dot{\theta}_B\cos\theta_B - c\dot{\theta}_C\cos\theta_C = 0.$$

If the crank angular velocity $\dot{\theta}_A$ is a known input, we are left with two equations and two unknowns, $\dot{\theta}_B$ and $\dot{\theta}_C$, so we can solve for them. A bit of algebra and trigonometric identities (including $\sin(A - B) = \sin A\cos B - \cos A\sin B$) are needed to simplify the angular velocity relationships into:

$$\dot{\theta}_B = \frac{a\dot{\theta}_A\sin(\theta_C - \theta_A)}{b\sin(\theta_B - \theta_C)}$$

$$\dot{\theta}_C = \frac{a\dot{\theta}_A\sin(\theta_A - \theta_B)}{c\sin(\theta_C - \theta_B)}.$$

The acceleration is similarly evaluated, noting that terms $\ddot{a}$, $\dot{a}$, $\ddot{b}$, $\dot{b}$, $\ddot{c}$, and $\dot{c}$ are zero:

$$\frac{d}{dt}\left[a\dot{\theta}_A i e^{i\theta_A} + b\dot{\theta}_B i e^{i\theta_B} - c\dot{\theta}_C i e^{i\theta_C}\right] = 0$$

$$a\ddot{\theta}_A i e^{i\theta_A} + a\dot{\theta}_A^2 i^2 e^{i\theta_A} + b\ddot{\theta}_B i e^{i\theta_B} + b\dot{\theta}_B^2 i^2 e^{i\theta_B} - c\ddot{\theta}_C i e^{i\theta_C} - c\dot{\theta}_C^2 i^2 e^{i\theta_C} = 0$$

$$a\ddot{\theta}_A i e^{i\theta_A} - a\dot{\theta}_A^2 e^{i\theta_A} + b\ddot{\theta}_B i e^{i\theta_B} - b\dot{\theta}_B^2 e^{i\theta_B} - c\ddot{\theta}_C i e^{i\theta_C} + c\dot{\theta}_C^2 e^{i\theta_C} = 0.$$

Again we substitute the Euler Equivalents Equation:

$$a\ddot{\theta}_A i \,(\cos\theta_A + i\sin\theta_A) - a\dot{\theta}_A^2 \,(\cos\theta_A + i\sin\theta_A) + b\ddot{\theta}_B i \,(\cos\theta_B + i\sin\theta_B)$$
$$- b\dot{\theta}_B^2 \,(\cos\theta_B + i\sin\theta_B) - c\ddot{\theta}_C i \,(\cos\theta_C + i\sin\theta_C) + c\dot{\theta}_C^2 \,(\cos\theta_C + i\sin\theta_C) = 0$$

$$a\ddot{\theta}_A i \cos\theta_A + a\ddot{\theta}_A i i \sin\theta_A - a\dot{\theta}_A^2 \cos\theta_A - a\dot{\theta}_A^2 i \sin\theta_A + b\ddot{\theta}_B i \cos\theta_B + b\ddot{\theta}_B i i \sin\theta_B$$
$$- b\dot{\theta}_B^2 \cos\theta_B - b\dot{\theta}_B^2 i \sin\theta_B - c\ddot{\theta}_C i \cos\theta_C - c\ddot{\theta}_C i i \sin\theta_C + \dot{\theta}_C^2 \cos\theta_C + c\dot{\theta}_C^2 i \sin\theta_C = 0$$

$$a\ddot{\theta}_A i \cos\theta_A - a\ddot{\theta}_A \sin\theta_A - a\dot{\theta}_A^2 \cos\theta_A - a\dot{\theta}_A^2 i \sin\theta_A + b\ddot{\theta}_B i \cos\theta_B$$
$$- b\ddot{\theta}_B \sin\theta_B - b\dot{\theta}_B^2 \cos\theta_B - b\dot{\theta}_B^2 i \sin\theta_B - c\ddot{\theta}_C i \cos\theta_C$$
$$+ c\ddot{\theta}_C \sin\theta_C + \dot{\theta}_C^2 \cos\theta_C + c\dot{\theta}_C^2 i \sin\theta_C = 0.$$

Separate real and imaginary terms and we have two equations:

$$- a\ddot{\theta}_A \sin\theta_A - a\dot{\theta}_A^2 \cos\theta_A - b\ddot{\theta}_B \sin\theta_B - b\dot{\theta}_B^2 \cos\theta_B + c\ddot{\theta}_C \sin\theta_C + c\dot{\theta}_C^2 \cos\theta_C = 0$$

$$a\ddot{\theta}_A \cos\theta_A - a\dot{\theta}_A^2 \sin\theta_A + b\ddot{\theta}_B \cos\theta_B - b\dot{\theta}_B^2 \sin\theta_B - c\ddot{\theta}_C \cos\theta_C + c\dot{\theta}_C^2 \sin\theta_C = 0.$$

Again, the input variable is often the acceleration of the crank, $\ddot{\theta}_A$, so the unknowns are $\ddot{\theta}_B$ and $\ddot{\theta}_C$. Solving these two equations for these unknowns yields:

$$\ddot{\theta}_B = \frac{a\ddot{\theta}_A \sin(\theta_A - \theta_C) + a\dot{\theta}_A^2 \cos(\theta_A - \theta_C) - c\dot{\theta}_C^2 + b\dot{\theta}_B^2 \cos(\theta_C - \theta_B)}{b\sin(\theta_C - \theta_B)}$$

$$\ddot{\theta}_C = \frac{a\ddot{\theta}_A \sin(\theta_A - \theta_B) + a\dot{\theta}_A^2 \cos(\theta_A - \theta_B) + b\dot{\theta}_B^2 - c\dot{\theta}_C^2 \cos(\theta_C - \theta_B)}{c\sin(\theta_C - \theta_B)}.$$

Again, the trigonometric identity $\sin(A - B) = \sin A \cos B - \cos A \sin B$ is needed to reduce down to this form.

Once these angular accelerations are known, the overall acceleration at any point can be found.

Example 17.2   (repeat of Examples 14.3, 15.3 and 16.3)

Crank BC of the slider-crank in Figure 17.10 rotates at a constant $\omega_{BC} = 8$ rad/s clockwise and is in position $\theta = 40°$. The dimensions are $L_{AB} = 500$ mm, $L_{BC} = 300$ mm, and $L_{BD} = 150$ mm. Determine the acceleration of Point A.

We match the variables given with those used in the previous sections:

$$a = L_{BC} = 300 \text{ mm}$$

$$b = L_{AB} = 500 \text{ mm}$$

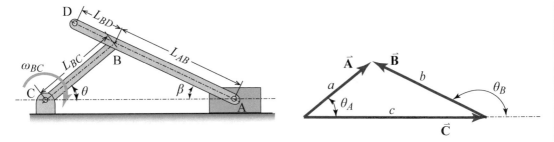

Figure 17.10: Example 17.2 (repeat of Figures 14.16, 15.12, and 16.11).

$$\theta_A = \theta = 40°.$$

Using the results from the position analysis we find:

$$\theta_B = \sin^{-1}\left[\frac{a}{b}\sin\theta_A\right] = \sin^{-1}\left[\frac{(300)}{(500)}\sin\left(40°\right)\right] = 22.69°, \ 112.7°.$$

This matches what we found in Example 14.3, we use the obtuse angle. We don't really need the next dimension but will find it anyway:

$$c = a\cos\theta_A + b\cdot\cos\theta_B = (300)\cos\left(40°\right) + (500)\cdot\cos(112.7) = 691.1 \text{ mm}$$

The angular velocity of link ABD is found from:

$$\dot{\theta}_B = \frac{a\cdot\cos\theta_A}{b\cdot\cos\theta_B}\dot{\theta}_A = \frac{(0.300)\cdot\cos(40°)}{(0.500)\cdot\cos(112.7)}(-8) = 3.985 \text{ rad/s}.$$

This too matches the previous example results. The positive result tells us it is counter-clockwise. We don't need the velocity for what is asked in this example, but we have solved it previously so we can compare:

$$\dot{c} = -a\cdot\dot{\theta}_A\sin\theta_A + b\cdot\dot{\theta}_B\sin\theta_B$$
$$= -(0.300)\cdot(-8)\sin\left(40°\right) + (0.500)\cdot(3.985)\sin(112.7) = 2.311 \text{ m/s}.$$

This also matches the results of Examples 14.3 and 15.3. We can find the unknown angular and translational accelerations from:

$$\ddot{\theta}_B = \frac{\left[a\ddot{\theta}_A\cos\theta_A - a\dot{\theta}_A^2\sin\theta_A + b\dot{\theta}_B^2\sin\theta_B\right]}{b\cos\theta_B}$$

$$= \frac{\left[(0.300)(0)\cos(40°) - (0.300)(-8)^2\sin(40°) + (0.500)(3.985)^2\sin(112.7)\right]}{(0.500)\cos(112.7)}$$

$$= -20.11 \text{ rad/s}^2$$

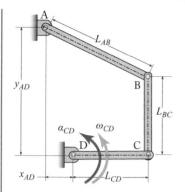
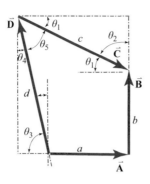
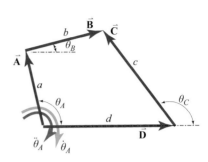

Figure 17.11: Example 17.3 including vector loops matching problem and reoriented to match Figure 17.8.

$$\ddot{c} = -a\ddot{\theta}_A \sin\theta_A - a\dot{\theta}_A^2 \cos\theta_A + b\ddot{\theta}_B \sin\theta_B + b\dot{\theta}_B^2 \cos\theta_B$$

$$= -(0.300)(0)\sin(40°) - (0.300)(-8)^2 \cos(40°) + (0.500)(-20.11)\sin(112.7)$$

$$+ (0.500)(3.985)^2 \cos(112.7) = 12.68 \text{ m/s.}$$

Both the angular acceleration of the connecting rod and the acceleration of the piston match the results of Example 16.3.

Answer: $\boxed{\vec{a}_A = 12.7 \text{ m/s}^2 \leftarrow}$.

Example 17.3   (repeat of Examples 14.4, 15.4, and 16.4)

In the position shown, bar *CD* of the four-bar mechanism in Figure 17.11 has an angular velocity of $\omega_{CD} = 12$ rad/s counter-clockwise and accelerating at $\alpha_{CD} = 25$ rad/s^2 counter-clockwise. The dimensions are $L_{AB} = 600$ mm, $L_{BC} = 400$ mm, $L_{CD} = 350$ mm, and $x_{AD} = 200$ mm. Determine the angular acceleration of Rocker *AB*.

Position Analysis:

The geometry of this problem doesn't match the four-bar vector loop setup we based the equation on, so we need to modify the orientation to match the variables. The vector loops shown in Figure 17.11 are labeled to coincide with the equations setup for Figure 17.8. The dimensions not provided are found from:

$$y_{AD} = L_{BC} + \sqrt{L_{AB}^2 - (L_{CD} + x_{AD})^2}$$

$$= (400) + \sqrt{(600)^2 - ((350) + (200))^2} = 639.9 \text{ mm}$$

$$\theta_1 = \sin^{-1}\left[\frac{y_{AD} - L_{BC}}{L_{AB}}\right] = \sin^{-1}\left[\frac{(639.9) - (400)}{(600)}\right] = 23.56°$$

$$\theta_2 = 90° - 23.56° = 66.44°$$

$$\theta_3 = \tan^{-1}\left[\frac{y_{AD}}{x_{AD}}\right] = \tan^{-1}\left[\frac{(639.9)}{(200)}\right] = 72.64°$$

$$\theta_4 = 90° - 72.64° = 17.36°$$

$$\theta_5 = 90° - 23.56° - 17.36° = 49.08°.$$

In order to match the vector loop equations we flip and rotate the links. Because we've flipped the image, the direction of the rotation also changes since we're looking at it from behind:

$$a = L_{CD} = 350 \text{ mm} \qquad \theta_A = 90° + 17.36° = 107.4°$$

$$b = L_{BC} = 400 \text{ mm} \qquad \theta_B = 17.36°$$

$$c = L_{AB} = 600 \text{ mm} \qquad \theta_C = 180° - 49.08° = 130.9°$$

$$d = \sqrt{(200)^2 + (400 + (600)\sin(23.56°))^2} = 670.4 \text{ mm}$$

$$\dot{\theta}_A = \omega_{CD} = -12 \text{ rad/s}$$

$$\ddot{\theta}_A = \alpha_{CD} = -25 \text{ rad/s}^2.$$

Velocity Analysis:

$$\dot{\theta}_B = \frac{a\dot{\theta}_A \sin(\theta_C - \theta_A)}{b\sin(\theta_B - \theta_C)} = \frac{(0.350)(-12)\sin((130.9°) - (107.4°))}{(0.400)\sin((17.36°) - (130.9°))} = 4.579 \text{ rad/s}$$

$$\dot{\theta}_C = \frac{a\dot{\theta}_A \sin(\theta_A - \theta_B)}{c\sin(\theta_C - \theta_B)} = \frac{(0.350)(-12)\sin((107.4°) - (17.36°))}{(0.600)\sin((130.9°) - (17.36°))} = -7.637 \text{ rad/s}$$

Acceleration Analysis:

$$\ddot{\theta}_B = \frac{a\ddot{\theta}_A \sin(\theta_A - \theta_C) + a\dot{\theta}_A^2 \cos(\theta_A - \theta_C) - c\dot{\theta}_C^2 + b\dot{\theta}_B^2 \cos(\theta_C - \theta_B)}{b\sin(\theta_C - \theta_B)}$$

$$= \frac{(0.350)(-25)\sin((107.4°) - (130.9°)) + (0.350)(-12)^2\cos((107.4°) - (130.9°))\,..}{(0.400)\sin((130.9°) - (17.36°))}$$

$$\frac{...- (0.600)(-7.637)^2 + (0.400)(4.579)^2\cos((130.9°) - (17.36°))}{(0.400)\sin((130.9°) - (17.36°))}$$

$$= 30.99 \text{ rad/s}^2$$

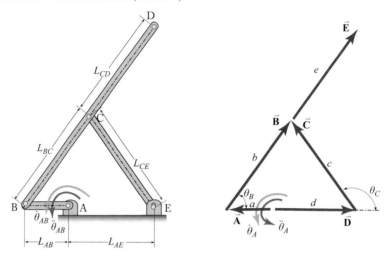

Figure 17.12: Hoeken straight line mechanism as a vector loop in Example 17.4.

$$\ddot{\theta}_C = \frac{a\ddot{\theta}_A \sin{(\theta_A - \theta_B)} + a\dot{\theta}_A^2 \cos{(\theta_A - \theta_B)} + b\dot{\theta}_B^2 - c\dot{\theta}_C^2 \cos{(\theta_C - \theta_B)}}{c\sin{(\theta_C - \theta_B)}}$$

$$= \frac{(0.350)(-25)\sin{((107.4°)-(17.36°))} + (0.350)(-12)^2 \cos{((107.4°)-(17.36°))} ..}{(0.600)\sin{((130.9°)-(17.36°))}}$$

$$\frac{... + (0.400)(4.579)^2 - (0.600)(-7.637)^2 \cos{((130.9°)-(17.36°))}}{(0.600)\sin{((130.9°)-(17.36°))}}$$

$$= 24.75 \text{ rad/s}^2.$$

The positive answer usually means counter-clockwise, but the picture was flipped to match the vector loop arrangement. So the answer is actually in the reverse direction, clockwise.

Answer: $\boxed{\ddot{\theta}_A = 24.8 \text{ rad/s}^2 \circlearrowright}$.

Example 17.4
Repeat Example 17.1 using the vector loop method.

Figure 17.12 shows the Hoeken Straight Line Mechanism represented as a vector loop. The dimensions are $L_{AB} = 1$ in, $L_{BC} = 2.5$ in, $L_{CD} = 2.5$ in, $L_{CE} = 2.5$ in, and $L_{AE} = 2$ in. Crank AB rotates counter-clockwise at $\dot{\theta}_{AB} = 10$ rad/s and $\ddot{\theta}_{AB} = 2$ rad/s^2. Determine the velocity and acceleration of Point D when crank AB is horizontal as shown. Note that we expect the velocity to be completely horizontal since this is the straight line portion:

$$a = L_{AB} = 1 \text{ in} \qquad \theta_A = 180°$$

$$b = L_{BC} = 2.5 \text{ in} \qquad \theta_B = 53.13°$$

$$c = L_{CE} = 2.5 \text{ in} \qquad \theta_C = 180° - 53.13° = 126.9°$$

$$d = L_{AE} = 2 \text{ in}$$

$$\dot{\theta}_A = \dot{\theta}_{AB} = 10 \text{ rad/s}$$

$$\ddot{\theta}_A = \ddot{\theta}_{AB} = 2 \text{ rad/s}^2$$

$$\dot{\theta}_B = \frac{a\dot{\theta}_A \sin(\theta_C - \theta_A)}{b \sin(\theta_B - \theta_C)} = \frac{(1)(10)\sin((126.9°) - (180°))}{(2.5)\sin((53.13°) - (126.9°))} = 3.333 \text{ rad/s}$$

$$\dot{\theta}_C = \frac{a\dot{\theta}_A \sin(\theta_A - \theta_B)}{c \sin(\theta_C - \theta_B)} = \frac{(1)(10)\sin((180°) - (53.13°))}{(2.5)\sin((126.9°) - (53.13°))} = 3.333 \text{ rad/s}$$

$$\ddot{\theta}_B = \frac{a\ddot{\theta}_A \sin(\theta_A - \theta_C) + a\dot{\theta}_A^2 \cos(\theta_A - \theta_C) - c\dot{\theta}_C^2 + b\dot{\theta}_B^2 \cos(\theta_C - \theta_B)}{b \sin(\theta_C - \theta_B)}$$

$$= \frac{(1)(2)\sin((180°) - (126.9°)) + (1)(10)^2 \cos((180°) - (126.9°)) ..}{(2.5)\sin((126.9°) - (53.13°))}$$

$$\frac{... - (2.5)(3.333)^2 + (2.5)(3.333)^2 \cos((126.9°) - (53.13°))}{(2.5)\sin((126.9°) - (53.13°))}$$

$$= 17.33 \text{ rad/s}^2$$

$$\frac{(1.600) + (60.00) - (27.77) + (7.762)}{(2.400)}.$$

We don't need to solve for the angular acceleration of rocker CE but will include it for comparison to Example 17.1:

$$\ddot{\theta}_C = \frac{a\ddot{\theta}_A \sin(\theta_A - \theta_B) + a\dot{\theta}_A^2 \cos(\theta_A - \theta_B) + b\dot{\theta}_B^2 - c\dot{\theta}_C^2 \cos(\theta_C - \theta_B)}{c \sin(\theta_C - \theta_B)}$$

$$= \frac{(1)(2)\sin((180°) - (53.13°)) + (1)(10)^2 \cos((180°) - (53.13°)) ..}{(2.5)\sin((126.9°) - (53.13°))}$$

$$\frac{... + (2.5)(3.333)^2 - (2.5)(3.333)^2 \cos((126.9°) - (53.13°))}{(2.5)\sin((126.9°) - (53.13°))}$$

$$= -16.00 \text{ rad/s}^2.$$

Both angular accelerations match what was found in Example 17.1. With the angular acceleration of the coupler we can find the acceleration at Point D once we know the acceleration of Point B as the reference:

$$\vec{a}_D = \vec{a}_B + (\vec{a}_{D/B})_t + (\vec{a}_{D/B})_n.$$

x-dir:

$$(a_D)_x = (a_B)_x + \left(\left(a_{D/B}\right)_t\right)_x + \left(\left(a_{D/B}\right)_n\right)_x$$

$$(a_D)_x = (a_B)_x - \ddot{\theta}_{BC}\left(L_{BC} + L_{BC}\right)\sin\theta_{BC} - \dot{\theta}_{BC}^2\left(L_{BC} + L_{BC}\right)\cos\theta_{BC}$$

$$(a_D)_x = (100) - (17.33)(5)\sin\left(53.13°\right) - (3.333)^2(5)\cos\left(53.13°\right) = -2.667 \text{ in/s}^2.$$

y-dir:

$$(a_D)_y = (a_B)_y + \left(\left(a_{D/B}\right)_t\right)_y + \left(\left(a_{D/B}\right)_n\right)_y$$

$$(a_D)_y = -(a_B)_y + \ddot{\theta}_{BC}\left(L_{BC} + L_{BC}\right)\cos\theta_{BC} - \dot{\theta}_{BC}^2\left(L_{BC} + L_{BC}\right)\sin\theta_{BC}$$

$$(a_D)_y = -(2) + (17.33)(5)\cos\left(53.13°\right) - (3.333)^2(5)\sin\left(53.13°\right) = 5.556 \text{ in/s}^2.$$

Answer: $\boxed{\vec{a}_D = (-2.67)\,\hat{\mathbf{i}} + (5.56)\,\hat{\mathbf{j}} \text{ in/s}^2}$.

This is the same result as in Example 17.1. The vector loop approach is well suited to solving more complicated mechanisms with multiple links (and loops) since it is more systematic than the previous approaches we have discussed. Upper-level engineering courses in mechanisms will often use this method.

REFERENCES:

Norton, R. L. 2020. *Design of Machinery*, 6th ed., McGraw-Hill.

Myszka, D. H. 2012. *Machines and Mechanisms: Applied Kinematic Analysis*, Pearson.

Uicker, J. J., Pennock, G. R., and Shigley, J. E. 2011. *Theory of Machines and Mechanisms*, vol. 1, Oxford University Press, New York.

Book 2 - Class 18   https://www.youtube.com/watch?v=BEOXA1QriIA

Coriolis Acceleration Analysis

B.L.U.F. (Bottom Line Up Front)

- Acceleration Analysis where the reference frame is also in motion creates an extra term that must be included.

- We've already introduced the "Coriolis" acceleration in Particle Polar Coordinates acceleration: $2\dot{r}\dot{\theta}$.

- The complete acceleration equation for planar rigid body motion is:

$$\vec{\mathbf{a}}_B = \vec{\mathbf{a}}_C + \ddot{\theta}_{BC}\hat{\mathbf{k}} \times \vec{\mathbf{r}}_{B/C} - \dot{\theta}_{BC}^2\,\vec{\mathbf{r}}_{B/C} + 2\dot{\theta}_{BC}\hat{\mathbf{k}} \times \vec{\mathbf{v}}_{B/C} + \vec{\mathbf{a}}_{B/C,rel}.$$

18.1   CORIOLIS ACCELERATION, SLIDING CONTACT, AND ROTATING FRAME ACCELERATION ANALYSIS

Recall for particles in polar coordinates the acceleration vector was written as:

$$\vec{\mathbf{a}} = \left(\ddot{r} - r\dot{\theta}^2\right)\hat{\mathbf{e}}_r + \left(r\ddot{\theta} + 2\dot{r}\dot{\theta}\right)\hat{\mathbf{e}}_\theta.$$

The multiple terms within the parentheses were the consequence of the unit vector changing orientation with time. The term that sticks out most is $2\dot{r}\dot{\theta}$ which we've previously introduced as "Coriolis Acceleration." It exists when there is both radial and angular velocity. Figure 18.1 shows an instance where Newtdog is walking outward ($\dot{r}$) on a Merry-Go-Round that is spinning ($\dot{\theta}$). One association we might make with this scenario is to think about the path Newtdog would need to travel in order to appear to move in a straight line to a stationary observer. His path would need to have a curve, and he would need to exert a force in order to make his path curve. This force can be associated with Coriolis acceleration.

A similar phenomenon occurs in rigid bodies when rotation and sliding action are both present in a linkage. The "rigid" in rigid body motion up until now assumed that the components' lengths don't change. But when there is sliding action a similar situation to the length change of a component exists. Figure 18.2 shows two scenarios where this is the case: an expanding link and a link with a sliding joint.

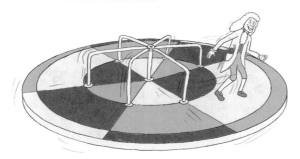

Figure 18.1: Newtdog experiencing Coriolis Acceleration on a merry-go-round (© E. Diehl).

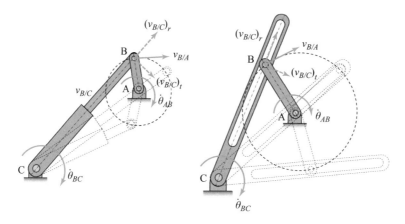

Figure 18.2: Links with changing length where Coriolis Acceleration exists.

The velocity of joint B is written as:

$$\vec{\mathbf{v}}_B = \vec{\mathbf{v}}_C + \vec{\mathbf{v}}_{B/C} = \vec{\mathbf{v}}_C + \underbrace{\dot{\theta}_{BC}\hat{\mathbf{k}} \times \vec{\mathbf{r}}_{B/C}}_{(v_{B/C})_t} + \underbrace{\vec{\mathbf{v}}_{B/C,rel}}_{(v_{B/C})_r}.$$

Here we recognize terms from previous velocity classes, namely a tangential relative motion component and a radial relative motion component. The basic relative acceleration concept still holds true for a rigid body:

$$\vec{\mathbf{a}}_B = \vec{\mathbf{a}}_C + \vec{\mathbf{a}}_{B/C}.$$

The relative motion of the joint is then represented by the complete planar rigid body acceleration equation written as:

$$\vec{\mathbf{a}}_B = \vec{\mathbf{a}}_C + \ddot{\theta}_{BC}\hat{\mathbf{k}} \times \vec{\mathbf{r}}_{B/C} - \dot{\theta}_{BC}^2 \vec{\mathbf{r}}_{B/C} + 2\dot{\theta}_{BC}\hat{\mathbf{k}} \times \vec{\mathbf{v}}_{B/C} + \vec{\mathbf{a}}_{B/C,rel}.$$

The term $\vec{\mathbf{a}}_{B/C}$ is included to address the acceleration of the changing length ($\ddot{r}$), and the $2\dot{\theta}_{BC}\hat{\mathbf{k}} \times \vec{\mathbf{v}}_{B/C}$ term represents Coriolis Acceleration. When trying to resolve the extra terms

Figure 18.3: Newtdog walking outward on a spinning playground merry-go-round in Example 18.1 (© E. Diehl).

in mechanisms we note that we can get the acceleration of joint B in the linkages in Figure 18.2 either from link AB or link BC. Therefore we can set the equations for acceleration from each equal to find those terms.

Recall from the previous class we can also use Complex Notation to represent links in Vector Loops. When the link length can change we represent the vector using the following equations with magnitude rate changes, in this case the terms $\dot{a}$ and $\ddot{a}$:

$$\vec{A} = a \cdot e^{i\theta_A}$$

$$\vec{v}_A = \dot{a}e^{i\theta_A} + a\dot{\theta}_A i e^{i\theta_A}$$

$$\vec{a}_A = \ddot{a}e^{i\theta_A} - a\dot{\theta}_A^2 e^{i\theta_A} + a\ddot{\theta}_A i e^{i\theta_A} + 2\dot{a}\dot{\theta}_A i e^{i\theta_A}.$$

We will show a simple example, followed by an example solved with vector math and then with vector loops, and finally with a long vector loop problem to demonstrate incorporating Coriolis Acceleration in rigid body kinematics.

Example 18.1
At the instant shown in Figure 18.3, Newtdog is moving outward on the spinning merry-go-round. When he is 4 ft from the center he is moving outward at a constant 1 ft/s. The merry-go-round is spinning at a constant 15 rpm clockwise. Determine the magnitude and direction of the acceleration.

We take stock of the variables given in the problem. We'll call Newtdog Point A and the center of the merry-go-round Point O:

$$\vec{r}_{A/O} = (4)\,\hat{i} \text{ ft}$$

$$\vec{v}_{A/O} = (1)\,\hat{i} \text{ ft/s}$$

$$\dot{\theta} = \frac{(-15 \text{ rpm})(2\pi \text{ rad/rev})}{(60 \text{ s/min})} = -1.571 \text{ rad/s}.$$

The general equation for planar rigid body acceleration applied to this scenario is:

$$\vec{\mathbf{a}}_A = \vec{\mathbf{a}}_O + \ddot{\theta}\hat{\mathbf{k}} \times \vec{\mathbf{r}}_{A/O} - \dot{\theta}^2 \vec{\mathbf{r}}_{A/O} + 2\dot{\theta}\hat{\mathbf{k}} \times \vec{\mathbf{v}}_{A/O} + \vec{\mathbf{a}}_{A/O,rel}.$$

Several of the terms are zero in this example. Since the pivot point of the merry-go-round is stationary $\vec{\mathbf{a}}_O = 0$, since it is rotating at a constant angular speed $\ddot{\theta}\hat{\mathbf{k}} \times \vec{\mathbf{r}}_{A/O} = 0$, and since Newtdog is walking at a constant outward rate $\vec{\mathbf{a}}_{A/O,rel} = 0$:

$$\vec{\mathbf{a}}_A = (0) + (0)\,\hat{\mathbf{k}} \times (4)\,\hat{\mathbf{i}} - (-1.571)^2\,(4)\,\hat{\mathbf{i}} + 2\,(-1.571)\,\hat{\mathbf{k}} \times (1)\,\hat{\mathbf{i}} + (0)$$

$$\vec{\mathbf{a}}_A = -(-1.571)^2\,(4)\,\hat{\mathbf{i}} + 2\,(-1.571)\,(1)\,\underbrace{\hat{\mathbf{k}} \times \hat{\mathbf{i}}}_{\hat{\jmath}}$$

$$\vec{\mathbf{a}}_A = (-9.870)\,\hat{\mathbf{i}} + (-3.142)\,\hat{\jmath}\ \text{ft/s}^2$$

$$a_A = \sqrt{(-9.870)^2 + (-3.142)^2} = 10.36\ \text{ft/s}^2$$

$$\theta_{A,a} = \tan^{-1}\left(\frac{(-3.142)}{(-9.870)}\right) = 17.66°\ \nearrow$$

$$\boxed{\vec{\mathbf{a}}_A = 10.4\ \text{ft/s}^2\ \ 17.7°\ \nearrow}.$$

One might say this isn't exactly a rigid body example. The merry-go-round is rigid but Newtdog is essentially a particle on the rigid body. This type of problem is often referred to as a "rotating frame" which is also an instance where Coriolis Acceleration arises.

Example 18.2
At the instant shown in Figure 18.4 link AB is at $\theta_{AB} = 60°$ and traveling $\dot{\theta}_{AB} = 3$ rad/s clockwise increasing by $\ddot{\theta}_{AB} = 5$ rad/s^2. Link AB is $L_{AB} = 1$ m long and the pin at B is located at $L_{BC} = 0.75$ m on slotted link BC. Determine the velocity and acceleration of the length in the slot and the angular speed and acceleration of slotted link BC:

$$\dot{\theta}_{AB} = (-3)\,\hat{\mathbf{k}}\ \text{rad/s}$$
$$\ddot{\theta}_{AB} = (-5)\,\hat{\mathbf{k}}\ \text{rad/s}^2$$

$$\vec{\mathbf{r}}_{B/A} = L_{AB}\cos\theta_{AB}\hat{\mathbf{i}} + L_{AB}\sin\theta_{AB}\hat{\mathbf{j}}$$
$$= (1)\cos\left(60°\right)\hat{\mathbf{i}} + (1)\sin\left(60°\right)\hat{\mathbf{j}} = (0.5)\,\hat{\mathbf{i}} + (0.8660)\,\hat{\mathbf{j}}\ \text{m}$$

$$\vec{\mathbf{r}}_{B/C} = L_{BC}\hat{\mathbf{i}} = (0.75)\,\hat{\mathbf{i}}\ \text{m}$$

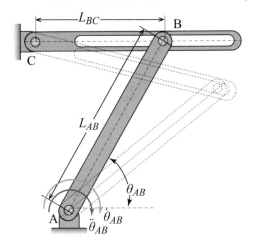

Figure 18.4: Links with a sliding joint in Example 18.2.

Velocity of joint B from link AB:

$$\vec{\mathbf{v}}_B = \vec{\mathbf{v}}_A + \vec{\mathbf{v}}_{B/A} = \vec{\mathbf{v}}_A + \dot{\theta}_{AB}\hat{\mathbf{k}} \times \vec{\mathbf{r}}_{B/A} = (0) + (-3)\,\hat{\mathbf{k}} \times \left[(0.5)\,\hat{\mathbf{i}} + (0.8660)\,\hat{\mathbf{j}} \right]$$

$$\vec{\mathbf{v}}_B = (-3)\,(0.5)\,\underbrace{\hat{\mathbf{k}} \times \hat{\mathbf{i}}}_{\hat{\mathbf{j}}} + (-3)\,(0.8660)\,\underbrace{\hat{\mathbf{k}} \times \hat{\mathbf{j}}}_{-\hat{\mathbf{i}}} = (2.598)\,\hat{\mathbf{i}} + (-1.500)\,\hat{\mathbf{j}}.$$

Velocity of joint B from link BC:

$$\vec{\mathbf{v}}_B = \vec{\mathbf{v}}_C + \vec{\mathbf{v}}_{B/C}$$

$$= \vec{\mathbf{v}}_C + \dot{\theta}_{BC}\hat{\mathbf{k}} \times \vec{\mathbf{r}}_{B/C} + \vec{\mathbf{v}}_{B/C,rel} = (0) + \dot{\theta}_{BC}\hat{\mathbf{k}} \times (0.75)\,\hat{\mathbf{i}} + v_{B/C,rel}\hat{\mathbf{i}}$$

$$\vec{\mathbf{v}}_B = \dot{\theta}_{BC}\,(0.75)\,\underbrace{\hat{\mathbf{k}} \times \hat{\mathbf{i}}}_{\hat{\mathbf{j}}} + v_{B/C,rel}\hat{\mathbf{i}} = v_{B/C,rel}\hat{\mathbf{i}} + \dot{\theta}_{BC}\,(0.75)\,\hat{\mathbf{j}}.$$

Equating the two expressions of the velocity of B:

$$(2.598)\,\hat{\mathbf{i}} + (-1.500)\,\hat{\mathbf{j}} = v_{B/C,rel}\hat{\mathbf{i}} + \dot{\theta}_{BC}\,(0.75)\,\hat{\mathbf{j}}$$

$$v_{B/C,rel} = 2.598 \text{ m/s } \rightarrow$$

$$(-1.500) = \dot{\theta}_{BC}(0.75)$$

$$\dot{\theta}_{BC} = -2.000 \text{ rad/s} = 2.000 \text{ rad/s } \circlearrowright \; .$$

Acceleration of joint B from link AB

$$\vec{\mathbf{a}}_B = \vec{\mathbf{a}}_A + \ddot{\theta}_{AB}\hat{\mathbf{k}} \times \vec{\mathbf{r}}_{B/A} - \dot{\theta}_{AB}^2\,\vec{\mathbf{r}}_{B/A} + 2\dot{\theta}_{AB}\hat{\mathbf{k}} \times \vec{\mathbf{v}}_{B/A} + \vec{\mathbf{a}}_{B/A,rel}$$

$$\vec{\mathbf{a}}_B = (0) + (-5)\,\hat{\mathbf{k}} \times \left[(0.5)\,\hat{\mathbf{i}} + (0.8660)\,\hat{\mathbf{j}}\right]$$

$$- (-3)^2 \left[(0.5)\,\hat{\mathbf{i}} + (0.8660)\,\hat{\mathbf{j}}\right] + 2\,(-3)\,\hat{\mathbf{k}} \times (0) + (0)$$

$$\vec{\mathbf{a}}_B = (-5)\,(0.5)\,\underbrace{\hat{\mathbf{k}} \times \hat{\mathbf{i}}}_{\hat{\mathbf{j}}} + (-5)\,(0.8660)\,\underbrace{\hat{\mathbf{k}} \times \hat{\mathbf{j}}}_{-\hat{\mathbf{i}}} - (-3)^2\,(0.5)\,\underbrace{\hat{\mathbf{k}} \times \hat{\mathbf{i}}}_{\hat{\mathbf{j}}} - (0.8660)\,\underbrace{\hat{\mathbf{k}} \times \hat{\mathbf{j}}}_{-\hat{\mathbf{i}}}$$

$$\vec{\mathbf{a}}_B = (-0.1699)\,\hat{\mathbf{i}} + (-10.29)\,\hat{\mathbf{j}}\ \text{m/s}^2.$$

Acceleration of joint B from link BC

$$\vec{\mathbf{a}}_B = \vec{\mathbf{a}}_C + \ddot{\theta}_{BC}\hat{\mathbf{k}} \times \vec{\mathbf{r}}_{B/C} - \dot{\theta}_{BC}^2\,\vec{\mathbf{r}}_{B/C} + 2\dot{\theta}_{BC}\hat{\mathbf{k}} \times \vec{\mathbf{v}}_{B/C} + \vec{\mathbf{a}}_{B/C,rel}$$

$$\vec{\mathbf{a}}_B = (0) + \ddot{\theta}_{BC}\hat{\mathbf{k}} \times \left[(0.75)\,\hat{\mathbf{i}}\right] - (-2.000)^2\,(0.75)\,\hat{\mathbf{i}}$$

$$+ 2\,(-2.000)\,\hat{\mathbf{k}} \times \left[(2.598)\,\hat{\mathbf{i}}\right] + a_{B/C,rel}\hat{\mathbf{i}}$$

$$\vec{\mathbf{a}}_B = \ddot{\theta}_{BC}\,(0.75)\,\underbrace{\hat{\mathbf{k}} \times \hat{\mathbf{i}}}_{\hat{\mathbf{j}}} - (-2.000)^2\,(0.75)\,\hat{\mathbf{i}} + 2\,(-2.000)\,(2.598)\,\underbrace{\hat{\mathbf{k}} \times \hat{\mathbf{i}}}_{\hat{\mathbf{j}}} + a_{B/C,rel}\hat{\mathbf{i}}$$

$$\vec{\mathbf{a}}_B = \left[-(-2.000)^2\,(0.75) + a_{B/C,rel}\right]\hat{\mathbf{i}} + \left[2\,(-2.000)\,(2.598) + \ddot{\theta}_{BC}\,(0.75)\right]\hat{\mathbf{j}}$$

$$\vec{\mathbf{a}}_B = \left[(-3.000) + a_{B/C,rel}\right]\hat{\mathbf{i}} + \left[(-10.39) + \ddot{\theta}_{BC}\,(0.75)\right]\hat{\mathbf{j}}.$$

Equating the two expressions of the acceleration of B:

$$(-0.1699)\,\hat{\mathbf{i}} + (-10.29)\,\hat{\mathbf{j}} = \left[(-3.000) + a_{B/C,rel}\right]\hat{\mathbf{i}} + \left[(-10.39) + \ddot{\theta}_{BC}\,(0.75)\right]$$

$$a_{B/C,rel} = (-0.1699) + (3.000) = 2.830\ \text{m/s}^2\ \rightarrow$$

$$\ddot{\theta}_{BC} = \frac{(-10.29) - (-10.39)}{(0.75)} = -0.1360\ \text{rad/s}^2 = 0.1360\ \text{rad/s}^2\ \circlearrowleft\ .$$

Answers:

$$\boxed{\vec{\mathbf{v}}_{B/C,rel} = 2.60\ \text{m/s}\ \rightarrow} \qquad \boxed{\vec{\mathbf{a}}_{B/C,rel} = 2.83\ \text{m/s}^2\ \rightarrow}$$

$$\boxed{\dot{\theta}_{BC} = 2.00\ \text{rad/s}\ \circlearrowleft} \qquad \boxed{\ddot{\theta}_{BC} = 0.136\ \text{rad/s}^2\ \circlearrowleft}\ .$$

Example 18.3

Repeat Example 18.2 using the complex notation vector loop method.

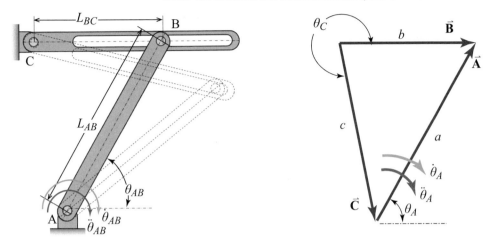

Figure 18.5: Vector loop of linkage in Example 18.3.

Figure 18.5 shows the vector loop representation of the linkage from Example 18.2. The parameters for the vector loop are:

$$a = 1 \text{ m} \qquad \theta_A = 60° \qquad \dot{\theta}_A = -3 \text{ rad/s}$$

$$\ddot{\theta}_A = -5 \text{ rad/s}^2 \qquad b = 1 \text{ m} \qquad \theta_B = 0°.$$

The dimensions of the "ground" link C aren't needed.

The vector loop equation is written as:

$$\vec{A} - \vec{B} + \vec{C} = 0 \qquad a \cdot e^{i\theta_A} - b \cdot e^{i\theta_B} + c \cdot e^{i\theta_C} = 0.$$

The derivatives are:

$$a\dot{\theta}_A i e^{i\theta_A} - \dot{b} e^{i\theta_B} - b\dot{\theta}_B i e^{i\theta_B} = 0$$

$$a\ddot{\theta}_A i e^{i\theta_A} - a\dot{\theta}_A^2 e^{i\theta_A} - b\ddot{\theta}_B i e^{i\theta_B} - \ddot{b} e^{i\theta_B} - 2\dot{b}\dot{\theta}_B i e^{i\theta_B} + b\dot{\theta}_B^2 e^{i\theta_B} = 0.$$

Apply the Euler Equivalent Equations to the velocity:

$$e^{i\theta_A} = \cos\theta_A + i\sin\theta_A$$

$$e^{i\theta_B} = \cos\theta_B + i\sin\theta_B$$

$$a\dot{\theta}_A i \left[\cos\theta_A + i\sin\theta_A\right] - \dot{b}\left[\cos\theta_B + i\sin\theta_B\right] - b\dot{\theta}_B i \left[\cos\theta_B + i\sin\theta_B\right] = 0$$

$$a\dot{\theta}_A i \cos\theta_A - a\dot{\theta}_A \sin\theta_A - \dot{b}\cos\theta_B - \dot{b}i\sin\theta_B - b\dot{\theta}_B i \cos\theta_B + b\dot{\theta}_B \sin\theta_B = 0.$$

Separating the real and imaginary terms, we get:

$$- a\dot{\theta}_A \sin\theta_A - \dot{b}\cos\theta_B + b\dot{\theta}_B \sin\theta_B = 0$$

$$a\dot\theta_A \cos\theta_A - \dot b \sin\theta_B - b\dot\theta_B \cos\theta_B = 0.$$

The two unknowns are $\dot\theta_B$ and $\dot b$.

$$-(1)(-3)\sin(60°) - \dot b \cos(0°) + (0.75)\dot\theta_B \sin(0°) = 0$$

$$(1)(-3)\cos(60°) - \dot b \sin(0°) - (0.75)\dot\theta_B \cos(0°) = 0$$

$$-(1)(-3)\sin(60°) - \dot b = 0$$

$$\dot b = 2.598 \text{ m/s}$$

$$(1)(-3)\cos(60°) - (0.75)\dot\theta_B = 0$$

$$\dot\theta_B = -2.000 \text{ rad/s}.$$

Apply the Euler Equivalent Equations to the acceleration:

$$e^{i\theta_A} = \cos\theta_A + i\sin\theta_A$$

$$e^{i\theta_B} = \cos\theta_B + i\sin\theta_B$$

$$a\ddot\theta_A i\,[\cos\theta_A + i\sin\theta_A] - a\dot\theta_A^2\,[\cos\theta_A + i\sin\theta_A] - b\ddot\theta_B i\,[\cos\theta_B + i\sin\theta_B]$$
$$- \ddot b\,[\cos\theta_B + i\sin\theta_B] - 2\dot b\dot\theta_B i\,[\cos\theta_B + i\sin\theta_B] + \dot\theta_B^2\,[\cos\theta_B + i\sin\theta_B] = 0$$

$$a\ddot\theta_A i\cos\theta_A - a\ddot\theta_A \sin\theta_A - a\dot\theta_A^2 \cos\theta_A - a\dot\theta_A^2 i\sin\theta_A - b\ddot\theta_B i\cos\theta_B + b\ddot\theta_B \sin\theta_B - \ddot b\cos\theta_B$$
$$+ \ddot b i\sin\theta_B - 2\dot b\dot\theta_B i\cos\theta_B + 2\dot b\dot\theta_B \sin\theta_B + b\dot\theta_B^2 \cos\theta_B + b\dot\theta_B^2 i\sin\theta_B = 0.$$

Separating the real and imaginary terms, we get:

$$-a\ddot\theta_A \sin\theta_A - a\dot\theta_A^2 \cos\theta_A + b\ddot\theta_B \sin\theta_B - \ddot b\cos\theta_B + 2\dot b\dot\theta_B \sin\theta_B + b\dot\theta_B^2 \cos\theta_B = 0$$

$$a\ddot\theta_A \cos\theta_A - a\dot\theta_A^2 \sin\theta_A - b\ddot\theta_B \cos\theta_B + \ddot b\sin\theta_B - 2\dot b\dot\theta_B \cos\theta_B + b\dot\theta_B^2 \sin\theta_B = 0.$$

The two unknowns are $\ddot\theta_A$ and $\ddot b$.

$$-(1)(-5)\sin(60°) - (1)(-3)^2 \cos(60°) + (0.75)\ddot\theta_B \sin(0°) - \ddot b\cos(0°)$$
$$+ 2(2.598)(-2.000)\sin(0°) + (0.75)(-2.000)^2 \cos(0°) = 0$$

$$(1)(-5)\cos(60°) - (1)(-3)^2 \sin(60°) - (0.75)\ddot\theta_B \cos(0°) + \ddot b\sin(0°)$$
$$- 2(2.598)(-2.000)\cos(0°) + (0.75)(-2.000)^2 \sin(0°) = 0$$

$$-(1)(-5)\sin(60°) - (1)(-3)^2 \cos(60°) - \ddot b + (0.75)(-2.000)^2 = 0$$

$$\ddot b = 2.830 \text{ m/s}^2$$

$$(1)(-5)\cos(60°) - (1)(-3)^2 \sin(60°) - (0.75)\ddot\theta_B - 2(2.598)(-2.000) = 0$$

$$\ddot\theta_B = -0.1360 \text{ rad/s}^2.$$

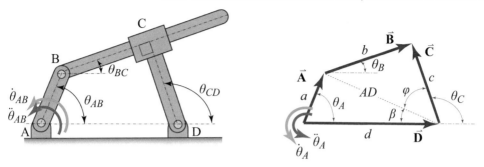

Figure 18.6: Inverter slider crank of Example 18.4.

Answers:

$$\vec{v}_{B/C,rel} = 2.60 \text{ m/s} \rightarrow$$

$$\vec{a}_{B/C,rel} = 2.83 \text{ m/s}^2 \rightarrow$$

$$\dot{\theta}_{BC} = 2.00 \text{ rad/s} \circlearrowleft$$

$$\ddot{\theta}_{BC} = 0.136 \text{ rad/s}^2 \circlearrowleft.$$

These are the same answers as in Example 18.2 which used Vector Math.

Example 18.4
The mechanism shown in Figure 18.6 is referred to as an "inverted slider crank." The parameters are $a = 2$ in, $\theta_A = 65°$, $c = 3$ in, and $d = 5$ in. Note that link BC and CD are perpendicular for this mechanism. The crank is traveling $\dot{\theta}_A = 10$ rad/s counter-clockwise and $\ddot{\theta}_A = 25$ rad/s^2 counter-clockwise. Determine the speed and acceleration magnitude of the length between B and C and the angular speed and acceleration of link BC.

The vector loop can be described using:

$$\vec{A} + \vec{B} - \vec{C} - \vec{D} = 0 \qquad a \cdot e^{i\theta_A} + b \cdot e^{i\theta_B} - c \cdot e^{i\theta_C} - d = 0 \quad \textcircled{1}.$$

Establish the relation $\theta_B + 90° = \theta_C$ ... taking the time derivative $\dot{\theta}_B = \dot{\theta}_C$ and again $\ddot{\theta}_B = \ddot{\theta}_C$.
Apply the Euler Equivalent Equations to the position:

$$e^{i\theta_A} = \cos\theta_A + i\sin\theta_A$$

$$e^{i\theta_B} = \cos\theta_B + i\sin\theta_B$$

$$e^{i\theta_C} = \cos\theta_C + i\sin\theta_C$$

$$a\cos\theta_A + ai\sin\theta_A + b\cos\theta_B + bi\sin\theta_B - c\cos\theta_C - ci\sin\theta_C - d = 0$$

$$a\cos\theta_A + b\cos\theta_B - c\cos(\theta_B + 90°) - d = 0$$

$$a \sin \theta_A + b \sin \theta_B - c \sin (\theta_B + 90°) = 0.$$

The unknowns are b and θ_B:

$$(2) \cos (65°) + b \cos \theta_B - (3) \cos (\theta_B + 90°) - (5) = 0$$

$$(2) \sin (65°) + b \sin \theta_B - (3) \sin (\theta_B + 90°) = 0$$

$$b \cos \theta_B - (3) \cos (\theta_B + 90°) - (4.155) = 0$$

$$(1.813) + b \sin \theta_B - (3) \sin (\theta_B + 90°) = 0.$$

While there are two equations and two unknowns, this is not an algebraically simple problem to solve. It's actually much easier to use the geometry. Drawing a line between joints A and D we can find the length using law of cosines:

$$AD = \sqrt{a^2 + d^2 - 2ad \cos \theta_A} = \sqrt{(2)^2 + (5)^2 - 2(2)(5)\cos (65°)} = 4.533 \text{ in}$$

$$AD = \sqrt{b^2 + c^2} \quad b = \sqrt{AD^2 - c^2} = \sqrt{(4.533)^2 - (3)^2} = 3.398 \text{ in.}$$

We can then find the angle between this line and link D we'll call β with the law of sines:

$$\frac{AD}{\sin \theta_A} = \frac{a}{\sin \beta}$$

$$\beta = \sin^{-1} \left[\frac{a}{AD} \sin \theta_A \right] = \sin^{-1} \left[\frac{(2)}{(4.533)} \sin (65°) \right] = 23.57°.$$

The angle between line AD and link C is:

$$\varphi = \cos^{-1} \left[\frac{c}{AD} \right] = \cos^{-1} \left[\frac{(3)}{(4.533)} \right] = 48.56°.$$

The angle of link C is

$$\theta_C = 180° - \beta - \varphi = 180° - (23.57°) - (48.56°) = 107.9°.$$

The angle of link B is

$$\theta_B = \theta_C - 90° = (107.9°) - 90° = 17.87°.$$

We can confirm this from the first of the equations we found from the vector loop:

$$b = \frac{(3) \cos ((17.87°) + 90°) + (4.155)}{\cos (17.87°)} = 3.398 \text{ in.}$$

We take the time derivative of equation (1):

$$a \dot{\theta}_A i e^{i\theta_A} + b \dot{\theta}_B i e^{i\theta_B} - \dot{b} e^{i\theta_B} - c \dot{\theta}_C i e^{i\theta_C} = 0 \quad \text{(2)}.$$

Substitute: $\dot{\theta}_C = \dot{\theta}_B$

$$a\dot{\theta}_A i e^{i\theta_A} + b\dot{\theta}_B i e^{i\theta_B} - \dot{b}e^{i\theta_B} - c\dot{\theta}_B i e^{i\theta_C} = 0.$$

Apply the Euler Equivalent Equations to the velocity:

$$a\dot{\theta}_A i \left[\cos\theta_A + i\sin\theta_A\right] + b\dot{\theta}_B i \left[\cos\theta_B + i\sin\theta_B\right]$$
$$- \dot{b}\left[\cos\theta_B + i\sin\theta_B\right] - c\dot{\theta}_B i \left[\cos\theta_C + i\sin\theta_C\right] = 0$$

$$a\dot{\theta}_A i \cos\theta_A - a\dot{\theta}_A \sin\theta_A + b\dot{\theta}_B i \cos\theta_B - b\dot{\theta}_B \sin\theta_B$$
$$- \dot{b}\cos\theta_B - \dot{b}i\sin\theta_B - c\dot{\theta}_B i \cos\theta_C + c\dot{\theta}_B \sin\theta_C = 0.$$

Separate the real and imaginary parts:

$$- a\dot{\theta}_A \sin\theta_A - b\dot{\theta}_B \sin\theta_B - \dot{b}\cos\theta_B + c\dot{\theta}_B \sin\theta_C = 0$$

$$a\dot{\theta}_A \cos\theta_A + b\dot{\theta}_B \cos\theta_B - \dot{b}\sin\theta_B - c\dot{\theta}_B \cos\theta_C = 0.$$

The two unknowns are $\dot{\theta}_B$ and $\dot{b}$. Entering in the known values:

$$- (2)(10)\sin(65°) - (3.398)\dot{\theta}_B \sin(17.87°) - \dot{b}\cos(17.87°) + (3)\dot{\theta}_B \sin(107.9°) = 0$$

$$- (15.27) - (1.043)\dot{\theta}_B - (0.9518)\dot{b} = 0$$

$$\dot{b} = \frac{-(15.27) - (1.043)\dot{\theta}_B}{(0.9518)} = -(16.04) - (1.096)\dot{\theta}_B$$

$$(2)(10)\cos(65°) + (3.398)\dot{\theta}_B \cos(17.87°) - \dot{b}\sin(17.87°) - (3)\dot{\theta}_B \cos(107.9°) = 0$$

$$(8.452) + (4.156)\dot{\theta}_B - (0.3069)\dot{b} = 0$$

$$\dot{b} = \frac{(8.452) + (4.156)\dot{\theta}_B}{(0.3069)} = (27.54) + (13.54)\dot{\theta}_B$$

$$- (16.04) - (1.096)\dot{\theta}_B = (27.54) + (13.54)\dot{\theta}_B$$

$$\dot{\theta}_B = -2.977 \text{ rad/s}$$

$$\dot{b} = -(16.04) - (1.096)(-2.977) = -12.78 \text{ in/s.}$$

Taking the time derivative of equation ② and simplifying, grouping terms:

$$a\ddot{\theta}_A i e^{i\theta_A} - a\dot{\theta}_A^2 e^{i\theta_A} + \ddot{\theta}_B i e^{i\theta_B} - b\dot{\theta}_B^2 e^{i\theta_B}$$
$$+ 2\dot{b}\dot{\theta}_B i e^{i\theta_B} + \ddot{b}e^{i\theta_B} - c\ddot{\theta}_C i e^{i\theta_C} - c\dot{\theta}_C^2 e^{i\theta_C} = 0.$$

Substitute: $\dot{\theta}_C = \dot{\theta}_B$ and $\ddot{\theta}_C = \ddot{\theta}_B$

$$a\ddot{\theta}_A i e^{i\theta_A} - a\dot{\theta}_A^2 e^{i\theta_A} + \ddot{\theta}_B i e^{i\theta_B} - b\dot{\theta}_B^2 e^{i\theta_B}$$
$$+ 2\dot{b}\dot{\theta}_B i e^{i\theta_B} + \ddot{b} e^{i\theta_B} - c\ddot{\theta}_B i e^{i\theta_C} - c\dot{\theta}_B^2 e^{i\theta_C} = 0.$$

Substitute the Euler Equivalent Equations and separate the real and imaginary parts

$$- a\ddot{\theta}_A \sin\theta_A - a\dot{\theta}_A^2 \cos\theta_A + b\ddot{\theta}_B \sin\theta_B + b\dot{\theta}_B^2 \cos\theta_B$$
$$+ 2\dot{b}\dot{\theta}_B \sin\theta_B - \ddot{b}\cos\theta_B + c\ddot{\theta}_B \sin\theta_C + c\dot{\theta}_B^2 \cos\theta_C = 0$$

$$a\ddot{\theta}_A \cos\theta_A - a\dot{\theta}_A^2 \sin\theta_A - b\ddot{\theta}_B \cos\theta_B + b\dot{\theta}_B^2 \sin\theta_B$$
$$- 2\dot{b}\dot{\theta}_B \cos\theta_B - \ddot{b}\sin\theta_B - c\ddot{\theta}_B \cos\theta_C + c\dot{\theta}_B^2 \sin\theta_C = 0.$$

The two unknowns are $\ddot{\theta}_B$ and $\ddot{b}$:

$$- (2)(25)\sin(65°) - (2)(10)^2 \cos(65°) + (3.398)\ddot{\theta}_B \sin(17.87°)$$
$$+ (3.398)(-2.977)^2 \cos(17.87°) + \underbrace{2(-12.78)(-2.977)\sin(17.87°)}_{Coriolis}$$
$$- \ddot{b}\cos(17.87°) + (3)\ddot{\theta}_B \sin(107.9°) + (3)(-2.977)^2 \cos(107.9°) = 0$$

$$- (45.32) - (84.52) + (1.043)\ddot{\theta}_B + (28.66)$$
$$+ \underbrace{(23.35)}_{Coriolis} - (0.9518)\ddot{b} + (2.855)\ddot{\theta}_B + (-8.172) = 0$$

$$- (86.00) + (3.898)\ddot{\theta}_B - (0.9518)\ddot{b} = 0$$

$$\ddot{b} = \frac{-(86.00) + (3.898)\ddot{\theta}_B}{(0.9518)} = -(90.36) + (4.095)\ddot{\theta}_B$$

$$(2)(25)\cos(65°) - (2)(10)^2 \sin(65°) - (3.398)\ddot{\theta}_B \cos(17.87°)$$
$$+ (3.398)(-2.977)^2 \sin(17.87°) - \underbrace{2(-12.78)(-2.977)\cos(17.87°)}_{Coriolis}$$
$$- \ddot{b}\sin(17.87°) - (3)\ddot{\theta}_B \cos(107.9°) + (3)(-2.977)^2 \sin(107.9°) = 0$$

$$(21.13) - (181.3) - (3.234)\ddot{\theta}_B + (9.241) - \underbrace{(72.42)}_{Coriolis}$$
$$- (0.3069)\ddot{b} + (0.9221)\ddot{\theta}_B + (25.30) = 0$$

$$- (198.0) - (2.312)\, \ddot{\theta}_B - (0.3069)\, \ddot{b} = 0$$

$$\ddot{b} = \frac{-(198.0) - (2.312)\, \ddot{\theta}_B}{(0.3069)} = -(645.2) - (7.533)\, \ddot{\theta}_B$$

$$-(90.36) + (4.095)\, \ddot{\theta}_B = -(645.2) - (7.533)\, \ddot{\theta}_B$$

$$\ddot{\theta}_B = -47.71 \ \text{rad/s}^2$$

$$\ddot{b} = -(645.2) - (7.533)\,(-47.71) = -285.8 \ \text{in/s}^2.$$

Answers:

$$\boxed{\dot{b} = -12.8 \ \text{in/s}} \qquad \boxed{\ddot{b} = -286 \ \text{in/s}^2}$$

$$\boxed{\dot{\theta}_{BC} = 2.98 \ \text{rad/s} \ \circlearrowleft} \qquad \boxed{\ddot{\theta}_{BC} = 286 \ \text{rad/s}^2 \ \circlearrowleft}.$$

This is a long problem with plenty of opportunities to make simple mistakes (especially with signs and sines). We can see the appeal of writing a computer code to solve this and more complicated problems. The benefit of working through a problem this way, however, is the insight we gain into the details of the contribution of each element, and especially that of Coriolis Acceleration.

Book 2 - Class 19

https://www.youtube.com/watch?v=EoNCPAofLYY

Mass Moment of Inertia

B.L.U.F. (Bottom Line Up Front)

- Kinetics of Rigid Bodies involves both translational and rotational influences.

- The center of mass is the location of an axis or axes with the smallest mass Moment of Inertia (mMoI) of an object.

- The mMoI resists rotational acceleration, just as mass resists translational acceleration.

- The mMoI describes the distribution of the mass in the rigid body and is the square of the distance of mass from an axis, just as Area Moment of Inertia is the square of the distance of area from an axis.

- We need the Parallel Axis Theorem (P.A.T.) to transfer the location of mMoI to an axis other than the one through the center of mass.

- It is very convenient to create a table when determining the center of mass and mMoI of an object by parts.

19.1   CENTER OF MASS

Rigid bodies have a point through which an axis (or axes in three dimensions) passes that it will rotate easiest, and therefore with the smallest mMoI (defined in Section 19.2, but for now think "least resistance to spin"). Before finding this smallest mMoI, we must first find the center of mass. In two-dimensions this is found from:

$$\overline{x} = \frac{\int x \, dm}{m_{total}} \qquad \overline{y} = \frac{\int y \, dm}{m_{total}},$$

where the distances x and y are from a convenient reference set of axes. Figure 19.1 shows Newtdog spinning a boomerang about its center of mass and the sketch of the boomerang we'd reference to find its location. Note that the x distance is measure from the y-axis and the y distance from the x-axis.

Often rigid bodies can be described by a combination of basic shapes such as rectangular or circular prisms either by addition or subtraction. In one plane the center of mass can be found

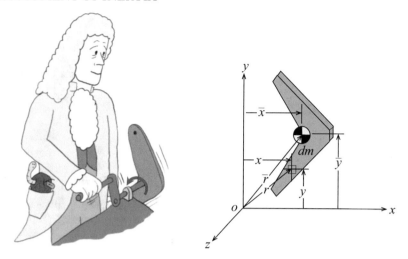

Figure 19.1: Newtdog spinning a boomerang about its center of mass (© E. Diehl).

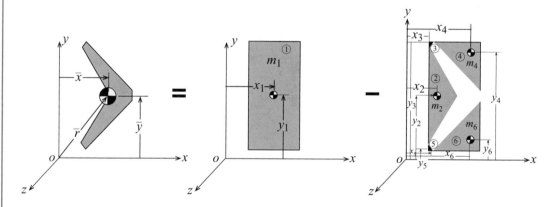

Figure 19.2: Finding the center of mass of the boomerang by subtracting triangles from a rectangle.

as a summation in the following equations:

$$\overline{x} = \frac{\sum x_i m_i}{\sum m_i} \qquad \overline{y} = \frac{\sum y_i m_i}{\sum m_i} \qquad \overline{r} = \sqrt{\overline{x}^2 + \overline{y}^2}.$$

When finding the center of mass by parts, we can add and/or subtract to form the desired shape. In Figure 19.2 we find the center of mass of the boomerang by subtracting triangles from an larger rectangle. This isn't the only way to approach finding this Center of Mass, since we could also have assembled it by adding smaller parts together.

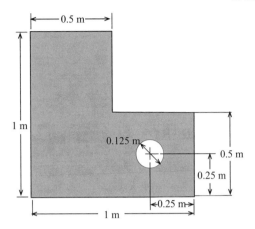

Figure 19.3: Example 19.1 center of mass.

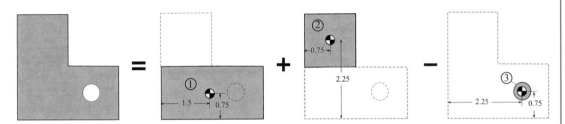

Figure 19.4: Example 19.1 breakdown of object into parts.

The Center of Mass is the same as the Center of Gravity but not always the same as the Centroid used in area calculations since the density and thickness isn't necessarily the same everywhere in a rigid body. For instance, a wooden object may have a lead weight in it, so the Center of Mass will be offset away from the centroid of the object. Since most of our problems deal with homogeneous objects in two dimensions, we often think of the centroid and Center of Mass as the same since under these conditions they are.

Example 19.1
Find the horizontal and vertical positions of the center of mass of the object in Figure 19.3 if the object is 0.5 m thick and has a density of 1,000 kg/m^3.

We need to break the object into parts and have a few choices to add or subtract basic shapes. One possible combination is shown in Figure 19.4. Here we choose to add a square above a horizontal rectangle, but we could just as easily subtract a smaller square from the upper-right corner of a larger square or added a square on the lower-right side to an upright rectangle.

Table 19.1: Example 19.1 center of mass

	r or x	y	z	d_x	d_y	d_z	V	m	md_x	md_y	md_z
	m	m	m	m	m	m	m^3	kg	kg*m	kg*m	kg*m
1	0.5	1	0.5	0.25	0.5	0.25	0.250	250.0	62.50	125.00	62.50
2	0.5	0.5	0.5	0.75	0.25	0.25	0.1250	125.0	93.75	31.25	31.25
3*	0.125		0.5	0.75	0.25	0.25	0.01227	-12.27	-9.204	-3.068	-3.068
							m_{total} =	362.73	147.0	153.2	90.7

	Centroids	
x_c	y_c	z_c
0.4054	0.4223	0.2500

The masses of the parts are found using density times volume:

$$m_1 = \rho V_1 = \rho b_1 h_1 t = (1000)\,(0.5)\,(1)\,(0.5) = 250.0 \text{ kg}$$

$$m_2 = \rho V_2 = \rho b_2 h_2 t = (1000)\,(0.5)\,(0.5)\,(0.5) = 125.0 \text{ kg}$$

$$m_3 = \rho V_3 = \rho \frac{\pi}{4} \rho d_3 t = (1000)\,\frac{\pi}{4}(0.125)^2\,(0.5) = 12.27 \text{ kg}.$$

The total mass is:

$$m_{total} = \sum m_i = m_1 + m_2 - m_3 = (250.0) + (125.0) - (12.27) = 362.7 \text{ kg}.$$

Using the left and bottom edges as references, we find the centers of mass from:

$$\bar{x} = \frac{\sum x_i m_i}{\sum m_i} = \frac{(250.0)\,(0.25) + (125.0)\,(0.75) - (12.27)\,(0.75)}{(362.7)} = \frac{(147.0)}{(362.7)} = 0.4054 \text{ m}$$

$$\bar{y} = \frac{\sum y_i m_i}{\sum m_i} = \frac{(250.0)\,(0.5) + (125.0)\,(0.25) - (12.27)\,(0.25)}{(362.7)} = \frac{(153.2)}{(362.7)} = 0.4223 \text{ m}.$$

This process lends itself to organizing it in Table 19.1. This is especially useful when done as a spreadsheet.

Answer: $\boxed{\bar{x} = 0.405 \text{ m}}$ $\boxed{\bar{y} = 0.422 \text{ m}}$.

19.2   MASS MOMENT OF INERTIA (mMoI)

The *distribution* of mass within a rigid body contributes to its resistance to rotational acceleration which is the mass Moment of Inertia (mMoI). "Distribution" is italicized because this is the

Figure 19.5: Newtdog on a swivel chair demonstrating mass moment of inertia (© E. Diehl).

operative word for how mMoI works. The definition of mMoI is:

$$\overline{I} = \int r^2 dm .$$

The distance, r, is squared which makes the mass further away from the axis especially important to the resulting resistance to rotation. Figure 19.5 shows Newtdog on a swivel chair with his arms tucked and outstretched. By extending his arms outward he significantly increases his mMoI, and therefore his resistance to spinning. We'll use this same concept when discussing angular momentum (specifically the conservation of angular momentum) in Class 23.

We are primarily dealing with two-dimensional objects in this text, but we should recognize there are multiple axes about which an object can spin. In two dimensions that axis is into the page. It's worth also considering the mMoIs when using the same two-dimensional object and giving it some thickness. In Figure 19.6 the boomerang spins about the three axes that pass through its center of mass. We will most often use the moment of inertia about the axis into the page (the z-axis in Figure 19.6).

The mMoI for each of the axes shown in Figure 19.6 can be found from:

$$\overline{I}_z = \int r^2 dm \qquad \overline{I}_x = \int y^2 dm \qquad \overline{I}_y = \int x^2 dm.$$

We are showing these as integration, but we'll typically find the mMoI from equations for common shapes or for more complex objects we'll add or subtract the mMoI of simple shapes as parts of the complex object. The basic shapes we use most often are summarized in Figure 19.7.

We might need other shapes than these, so searching for a reliable on-line reference is often necessary, but as always we'd want to be cautious of the legitimacy of the information.

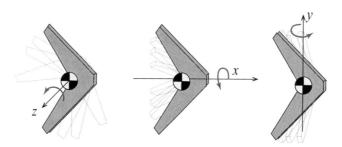

Figure 19.6: Rotation of boomerang about axes through center or mass.

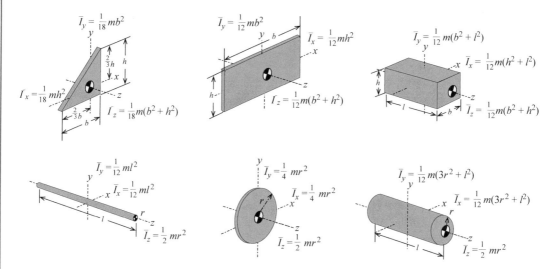

Figure 19.7: Mass moment of inertia of some basic shapes.

Another option is to use solid modeling software which will return the center of mass and the mMoI about it and other reference axes. It's good practice to use several methods when possible to confirm the results of hand calculations with computer results, and vice versa.

19.3   PARALLEL AXIS THEOREM (P.A.T.)

It is more difficult to rotate an object at an axis away from the center of mass, so when the axis of rotation moves away the "Parallel Axis Theorem" (P.A.T.) must be employed to account for the increased resistance to rotation. P.A.T. is written as:

$$\boxed{I = \overline{I} + md^2}.$$

The term d is the distance the axis is moved in a straight line. Figure 19.8 shows some examples of the boomerang rotating about non-centroidal axes in three dimensions. Picture the

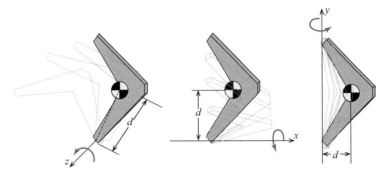

Figure 19.8: Rotation of boomerang about axes away from center of mass.

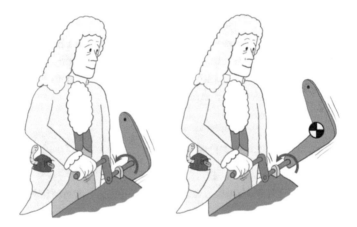

Figure 19.9: Newtdog spins the boomerang about is centroid and an axis at the tip (© E. Diehl).

increased difficulty of spinning each boomerang compared to the rotation of each in Figure 19.6. Figure 19.9 shows Newtdog testing this out on a tabletop rig.

P.A.T. further emphasizes that the contribution of mass farther away from the center of rotation to the MoI is much more significant than mass closer to it. This is especially true because the distance between axes is squared. The P.A.T. is also used when calculating the area moment of inertia, except that area is used instead of mass. While mMoI resists angular acceleration, area moment of inertia resists bending. Another similarity is that area located further from the neutral axis has much more effect on the value as demonstrated with I-beams.

It is often necessary to use P.A.T. multiple times when finding the mMoI by parts as each must be moved to a common axis before they can be added or subtracted.

19.4   RADIUS OF GYRATION

A useful parameter to know about a rigid body is the Radius of Gyration, found from $k_x = \sqrt{\frac{I_x}{m}}$.
It is often expressed as either "k" or "r" with a subscript to indicate about which axis it describes.
The Radius of Gyration is a "sort of average" of the mass distribution and for objects with the
same mass can express the way that mass is distributed in the body, closer or further away from
the center of mass. You might be given the radius of gyration as a way of communicating the
mMoI which you'd find from $I = k^2m$. The radius of gyration is also a parameter used when
discussing the area moment of inertia. These radii of gyration share the same concept (except
area is used rather than mass) but aren't numerically related, since the axis about which they refer
has isn't used the same way in the moment of inertia formulation.

19.5   mMoI BY INTEGRATION

The mMoI of less commonly used shaped objects can be found from integration. A cone is used
in Examples 19.2 and 19.3 in two different integration strategies even though we could look up
the mMoI. The desired mMoI in these examples is about the y axis so either rings or disks are
used as the differential mass element to integrate.

Example 19.2
Find a formula for the mMoI about the y-axis for the cone shown in Figure 19.10 using integration with hollow cylinders.

As shown in Figure 19.11, we break the cone down into concentric hollow cylinders with
thickness of dx.

The incremental volume of each hollow cylinder is found from the circumference of an
average circle in the cylinder wall ($2\pi x$) times the thickness (dx) times the height of the individual cylinder ($h - y$). We can relate the position by $y = \frac{h}{a}x$. The incremental mass is the
density times this incremental volume:

$$dm = \rho dV = \rho 2\pi x \left(h - \frac{h}{a}x \right) dx$$

$$m = \int_0^a \rho 2\pi x \left(h - \frac{h}{a}x \right) dx = \rho 2\pi \int_0^a \left(xh - \frac{h}{a}x^2 \right) dx$$

$$= \rho 2\pi \left(\frac{1}{2}a^2h - \frac{1}{3}\frac{h}{a}a^3 \right) = \frac{1}{3}\rho\pi a^2h$$

$$\bar{I}_y = \int x^2 dm$$

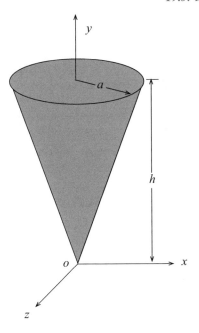

Figure 19.10: Example 19.2 cone.

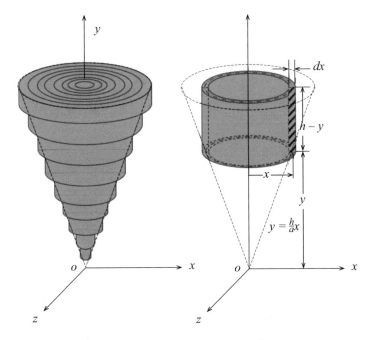

Figure 19.11: Example 19.2 cone broken into hollow cylinders.

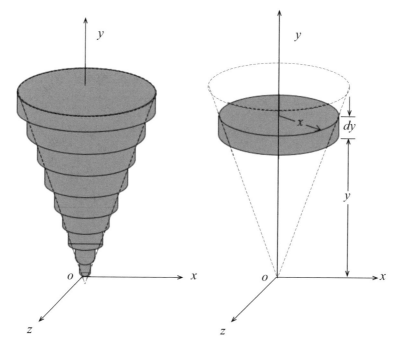

Figure 19.12: Example 19.3 cone broken into disks.

$$\overline{I}_y = \int_0^a x^2 \rho 2\pi x \left(h - \frac{h}{a}x\right) dx = \rho 2\pi \int_0^a \left(x^3 h - \frac{h}{a}x^4\right)$$

$$dx = \rho 2\pi \left(\frac{1}{4}a^4 h - \frac{1}{5}\frac{h}{a}a^5\right) = \frac{1}{10}\rho\pi a^4 h.$$

We can replace $\rho\pi a^2 h = 3m$.

Answer: $\boxed{\overline{I}_y = \frac{3}{10}ma^2}$.

Example 19.3
Find the formula for the mMoI about the y-axis of the cone shown in Figure 19.10 using integration with rings.

In order to use disks, we must assume we know the mMoI incremental disk to be $d\overline{I}_y = \frac{1}{2}r^2 dm$. We break the cone into disks as shown in Figure 19.12.

The incremental volume of each disk is found from the area (πx^2) times the thickness (dy). We can relate the position as a function of y using $x = \frac{a}{h}y$. The incremental mass is the density times this incremental volume:

$$dm = \rho dV = \rho\pi \left(\frac{a}{h}y\right)^2 dy$$

$$m = \int_0^h \rho\pi \frac{a^2}{h^2} y^2 dy = \frac{1}{3}\rho\pi a^2 h \quad \text{(which is the same as Example 19.1)}$$

$$\overline{I}_y = \int d\overline{I}_y = \int \frac{1}{2}r^2 dm = \int_0^h \frac{1}{2}\left(\frac{a}{h}y\right)^2 \rho\pi \left(\frac{a}{h}y\right)^2 dy$$

$$\overline{I}_y = \frac{1}{2}\rho\pi \left(\frac{a}{h}\right)^4 \int_0^h y^4 dy = \frac{1}{10}\rho\pi \left(\frac{a}{h}\right)^4 h^5 = \frac{1}{10}\rho\pi a^4 h.$$

We can replace $\rho\pi a^2 h = 3m$.

Answer: $\boxed{\overline{I}_y = \frac{3}{10}ma^2}$ which is the same as we found in Example 19.2.

19.6   mMoI FROM PARTS USING A TABLE

It is common to breakdown more complicated parts into the basic shapes of Figure 19.7, adding or subtracting them after they've been moved (via the P.A.T.) to a common axis from their individual center of mass axis. The general procedure by parts is either to write out the equations or to create a table to keep track of all of the pieces. In either case we must first find the center of mass, which can also benefit from being calculated within a table. Example 19.4 demonstrates this process.

Example 19.4
Determine the mMoI (in kg-m²) and the radii of gyration of the plastic part shown in Figure 19.13 with respect to the z-axis and the axis passing through the center of mass that is parallel to the z-axis. The density of this plastic is 1,000 kg/m³.

The center of mass of this object was found in Example 19.1 to be at $\overline{x} = 0.405$ m and $\overline{y} = 0.422$ m. Finding the center of mass is usually the first step in solving many mMoI problems. The masses of the individual parts and total were also found to be: $m_1 = 250.0$ kg, $m_2 = 125.0$ kg, $m_3 = 12.27$ kg, and $m_{total} = 362.7$ kg.

We can write this out as one long equation using the parts and variables shown in Figure 19.13:

$$I_z = \overline{I}_{z,1} + m_1 d_1^2 + \overline{I}_{z,2} + m_2 d_2^2 - \left(\overline{I}_{z,3} + m_3 d_3^2\right).$$

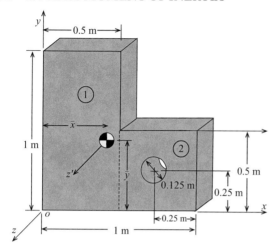

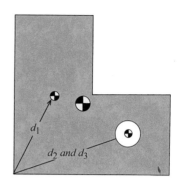

Figure 19.13: Example 19.4 mMoI.

$$I_z = \frac{1}{12}(250.0)\left((0.5)^2 + (1)^2\right) + (250.0)\left(\sqrt{(0.25)^2 + (0.5)^2}\right)^2$$

$$+ \frac{1}{12}(125.0)\left((0.5)^2 + (0.5)^2\right) + (125.0)\left(\sqrt{(0.75)^2 + (0.25)^2}\right)^2$$

$$- \left(\frac{1}{2}(12.27)(0.125)^2 + (12.27)\left(\sqrt{(0.75)^2 + (0.25)^2}\right)^2\right)$$

$$= 179.7 \text{ kg} \cdot \text{m}^2.$$

The radius of gyration about the z-axis is:

$$k_z = \sqrt{\frac{I_z}{m}} = \sqrt{\frac{(179.7)}{(362.7)}} = 3.995 \text{ m}.$$

The mMoI about the center of mass can be found by moving the axis using P.A.T. Noting that we must subtract because the mMoI will be less at the center of mass:

$$\overline{I}_z = I_z - m_{total}\left(\overline{x}^2 + \overline{y}^2\right) = (179.7) - (362.7)\left((0.4054)^2 + (0.4223)^2\right) = 55.44 \text{ kg} \cdot \text{m}^2.$$

The radius of gyration about the center of mass is:

$$k_{\overline{z}} = \sqrt{\frac{\overline{I}_z}{m}} = \sqrt{\frac{(55.44)}{(362.7)}} = 2.220 \text{ m}.$$

Table 19.2 shows this same calculation within a table. This lends itself to creating a spreadsheet which is especially useful when we want to find the mMoI of similarly shaped objects, or to design objects or to confirm our hand calculations.

Table 19.2: Example 19.4 mMol calculation

	r or x	y	m	lz	d_x	d_y	d	$l_z + m(d)^2$
	m	m	kg	kg*m²	m	m	m	kg*m²
1	0.5	1	250.0	26.0	0.25	0.5	0.55902	104.2
2	0.5	0.5	125.0	5.21	0.75	0.25	0.79057	83.33
3*	0.125	0	-12.272	-0.01598	0.75	0.25	0.79057	-7.686
							$l_z =$	179.8
							$d_{z-zc} =$	0.5854
							$l_{zc} =$	55.51

Answers:

$$\boxed{I_z = 0.4160 \text{ slug} \cdot \text{ft}^2} \qquad \boxed{k_z = 0.5939 \text{ ft}}$$

$$\boxed{\overline{I}_z = 0.1316 \text{ slug} \cdot \text{ft}^2} \qquad \boxed{k_z = 0.3340 \text{ ft}} \ .$$

Book 2 - Class 20

https://www.youtube.com/watch?v=9R6_VNV9i7c

C L A S S   20

Newton's Second Law in Constrained Plane Motion

B.L.U.F. (Bottom Line Up Front)

- Rigid Body Motion Kinetics has both translation and rotation.

- N2L for rigid bodies includes $\sum \vec{\mathbf{F}} = m\vec{\mathbf{a}}$ and $\sum \vec{\mathbf{M}}_{\mathbf{G}} = I_G \vec{\alpha}$.

- When the rotation is about a point other than the center of mass, we use "effective inertial moment" $\sum \vec{\mathbf{M}} = \sum \left(\vec{\mathbf{M}}\right)_{eff}$.

- The effective inertial moment concept demonstrates how the Parallel Axis Theorem works.

So far we've covered Kinematics of both particles (vol. 1) and rigid bodies and Kinetics of particles (vol. 1). Now we complete the four parts by introducing Kinetics of rigid bodies. Just as with Kinetics of particles we break the topic into three parts: Newton's 2nd Law (N2L), Work-Energy, and Impulse-Momentum. In the previous class we introduced the mMoI which, for rigid body rotation, is analogous to mass in particle translation. That is, mass is to translation as mMoI is to rotation.

20.1   N2L APPLIED TO A RIGID BODY

In Class 6 (vol. 1) we introduced Newton's 2nd Law for translation: $\sum \vec{\mathbf{F}} = m\vec{\mathbf{a}}$.

This remains true for rigid bodies, but we use the center of mass as the reference for the translational acceleration as if it were a particle at the center. In rigid body kinetics, the shape of the object is such that we must consider its rotation to fully define its motion. To account for the rotation of the rigid body we write N2L as $\sum \vec{\mathbf{M}}_{\mathbf{G}} = I_G \vec{\alpha}$ where "G" is the center of mass (therefore $I_G = \bar{I}$, which we'll use interchangeably).

In Figure 20.1 we draw the FBD/IBD pair again and include both the translational and rotational external forces and moments on the FBD (including reactions at supports) and the inertial forces and "effective inertial moments" on the IBD. We'll explain the need to refer to these as "effective inertial moments" this way in a little bit.

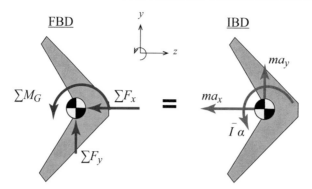

Figure 20.1: Free body diagram and inertial body diagram for rigid body kinetics.

Below the FBD/IBD pair we write out N2L in translation and rotation to emphasize that the diagrams match the equations.

$$\sum \vec{\mathbf{F}} = m\,\vec{\mathbf{a}}$$

$$\sum \vec{\mathbf{M_G}} = \bar{I}\,\vec{\boldsymbol{\alpha}}.$$

Recall that another way to describe N2L is the time rate change of momentum. Where linear momentum is written as $\vec{\mathbf{L}} = m\,\vec{\mathbf{v}}$, we introduce angular momentum about the center of mass as $\vec{\mathbf{H_G}} = I_G\,\vec{\boldsymbol{\omega}}$.

N2L for translation as the time rate change of momentum is:

$$\sum \vec{\mathbf{F}} = \frac{d\,\vec{\mathbf{L}}}{dt} = m\,\dot{\vec{\mathbf{v}}} = m\,\vec{\mathbf{a}}.$$

Now we have N2L for rotation as the time rate change of angular momentum:

$$\sum \vec{\mathbf{M_G}} = \frac{d\,\vec{\mathbf{H_G}}}{dt} = I_G\,\vec{\boldsymbol{\alpha}}.$$

This is a good concept to keep in mind in order to understand the relationship between rotational N2L and Angular Impulse-Momentum later on.

20.2 N2L OF RIGID BODIES CONSTRAINED ABOUT A FIXED POINT AND THE EFFECTIVE INERTIAL MOMENT

Recall from Statics that we can apply the summation of moments about various points, often strategically so we can omit reaction forces and reduce the unknowns. We introduce the idea

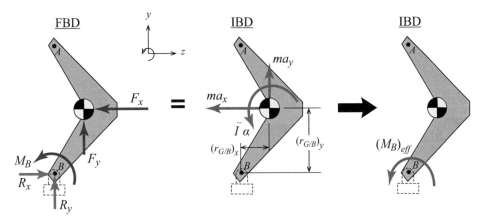

Figure 20.2: FBD/IBD for rigid body about a fixed point and the effective inertial mass.

of the "effective inertial moment" to account for dynamic rotation about an axis other than the center of mass (center of gravity). In Figure 20.2 we draw the FBD/IBD pair for the special case of motion about fixed Point B. An applied moment around the axis as well as forces on the center of gravity cause motion in both translation of the center of gravity and rotation. In order to remove the reaction forces at the pin we use it as the reference for the sum of the moments, but the inertial forces are away from the reference axis and cause an effective moment about the pin. This is why we treat them as an effective inertial moment, which in this case is written as: $\circlearrowleft \sum \left(\vec{\mathbf{M}}_B\right)_{eff} = \bar{I}\alpha + ma_y \left(r_{G/B}\right)_x + ma_x \left(r_{G/B}\right)_y$. Notice the sign convention for moment is positive for counter-clockwise rotation as usual.

Because the reference point for the sum of the moments is not always the center of gravity, we will use the following more general equations for application of N2L to rigid body motion:

$$\boxed{\sum \vec{\mathbf{F}} = m\,\vec{\mathbf{a}}} \qquad \boxed{\sum \vec{\mathbf{M}}_O = \sum \left(\vec{\mathbf{M}}_O\right)_{eff}}.$$

Recognize for the special case of rotation about a fixed axis the rotational N2L equation becomes: $\sum \vec{\mathbf{M}}_O = I_O \vec{\alpha}$, where O is the fixed point.

We will see in the next section that the parallel axis theorem (P.A.T.) can be derived from this effective moment concept.

20.3   P.A.T. REVEALED WHEN USING RIGID BODY N2L ABOUT A FIXED POINT

We must remember that the principles of kinematics remain a tool at our disposal when performing kinetics analysis. This is relevant when discussing P.A.T. because the magnitude of tangential acceleration in fixed axis rotation is kinematically related to the angular acceleration

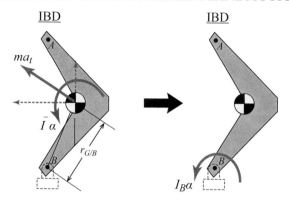

Figure 20.3: IBD for effective inertial moment about a fixed axis.

by: $a_t = \alpha r_{G/B}$. This can be applied to the effective inertial moment of the IBD in Figure 20.2, and as shown in Figure 20.3, will result in $\circlearrowleft \sum (\vec{M_B})_{eff} = \bar{I}\alpha + ma_t r_{G/B} = I_B\alpha$.

If we substitute $a_t = \alpha r_{G/B}$ we find $\bar{I}\alpha + mr_{G/B}{}^2\alpha = I_B\alpha$, and recognize $I_B = \bar{I} + mr_{G/B}{}^2$ is the P.A.T. This confirms the use of P.A.T. but also helps reinforce the need and usefulness of the effective inertial moment when applying N2L to rigid body motion.

We will similarly confirm the P.A.T. again when we discuss rigid body Work-Energy and Impulse-Momentum.

Example 20.1

Just like Example 5.3 (vol. 1), the three weights and pulley setups shown begin at rest. The difference in this arrangement is a single frictionless pulley weighing $W_C = 50$ lb, of radius $r_C = 1$ ft and a radius of gyration of $k_C = 0.75$ ft. In setup (a) the force ($P = 50$ lb) is applied to the cable attached to block A ($W_A = 75$ lb). Setup (b) has the same block A and is connected via a cable over the pulley to block B ($W_B = 50$ lb). Setup (c) has larger blocks with the same weight difference between them ($W_A = 175$ lb and $W_B = 150$ lb). Determine the acceleration of block A for each setup (Figure 20.4).

The mMoI of the pulley is found using the provided radius of gyration:

$$I_C = \frac{W_C}{g}k_C{}^2 = \frac{(50)}{(32.2)}(0.75)^2 = 0.8734 \text{ slug} \cdot \text{ft}^2.$$

(a) In the first scenario we draw the FBD/IBD pair in Figure 20.5.
Apply N2L

$$\circlearrowleft \sum M_C = \sum (M_C)_{eff}$$
$$W_A r_C - P r_C = I_C\alpha + m_A a_{Ay} r_C.$$

Recognize that $a_{Ay} = \alpha r_C$

$$W_A r_C - P r_C = I_C\alpha + m_A r_C{}^2\alpha$$

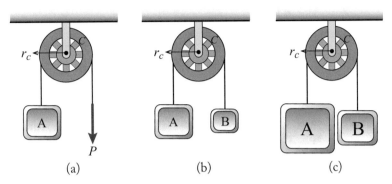

Figure 20.4: Example 20.1 mass and pulley problem.

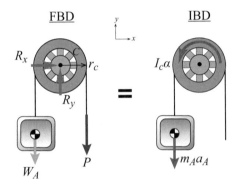

Figure 20.5: FBD/IBD of Example 20.1(a).

$$(75)\,(1) - (50)\,(1) = (0.8734)\,\alpha + \frac{(75)}{32.2}\,(1)^2\,\alpha$$

$$\alpha = 7.806 \text{ rad/s}^2$$

$$a_{Ay} = \alpha r_C = (7.806)\,(1) = 7.806 \text{ ft/s}^2$$

$$(a)\quad \boxed{\vec{a}_A = 7.81 \text{ ft/s}^2 \downarrow}.$$

Compare this to the results from Example 5.3 (vol. 1) of 10.73 ft/s², and we see the inclusion of a large pulley reduces the acceleration by 27%. We can see why many problems that include pulleys include the disclaimer they are massless. The remainder of the problem is included here for completeness, but the point of this example has been made. Note that the P.A.T. observation we've made doesn't really apply in this situation since the block is not actually part of the pulley, but we will see it demonstrated in other examples. We were able to treat the weight and the pulley together because we took the sum of the moments about the pin, allowing

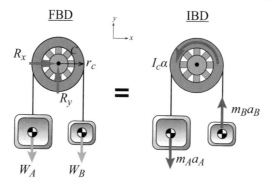

Figure 20.6: FBD/IBD of Example 20.1(b).

us to avoid the reaction forces of the pin. In Example 5.3 (vol. 1) we had to treat the weights separately in order to avoid them.

Part (b): The second scenario FBD/IBD pair is shown in Figure 20.6.
Apply N2L

$$\circlearrowleft \sum M_C = \sum (M_C)_{\textit{eff}}$$
$$W_A r_C - W_B r_C = I_C \alpha + m_A a_A r_C + m_B a_B r_C.$$

Recognize that $a_A = a_B = \alpha r_C$

$$W_A r_C - W_B r_C = I_C \alpha + m_A r_C{}^2 \alpha + m_B r_C{}^2 \alpha$$

$$(75)\,(1) - (50)\,(1) = (0.8734)\,\alpha + \frac{(75)}{32.2}(1)^2 \alpha + \frac{(50)}{32.2}(1)^2 \alpha$$

$$\alpha = 5.257 \text{ rad/s}^2$$

$$a_A = \alpha r_C = (5.257)\,(1) = 5.257 \text{ ft/s}^2.$$

(b) $\boxed{\vec{\mathbf{a}}_A = 5.26 \text{ ft/s}^2 \downarrow}$ 18% less than with a massless pulley.

Part (c): The third scenario FBD/IBD pair is shown in Figure 20.7.
Apply N2L

$$\circlearrowleft \sum M_C = \sum (M_C)_{\textit{eff}}$$
$$W_A r_C - W_B r_C = I_C \alpha + m_A a_A r_C + m_B a_B r_C.$$

Recognize that $a_A = a_B = \alpha r_C$

$$W_A r_C - W_B r_C = I_C \alpha + m_A r_C^2 \alpha + m_B r_C^2 \alpha$$

$$(175)\,(1) - (150)\,(1) = (0.8734)\,\alpha + \frac{(175)}{32.2}(1)^2 \alpha + \frac{(150)}{32.2}(1)^2 \alpha$$

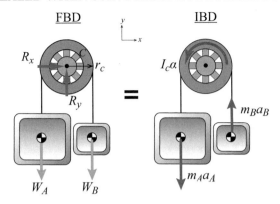

Figure 20.7: FBD/IBD of Example 20.1(c).

Figure 20.8: Newtdog on a beam in Example 20.2 (© E. Diehl).

$$\alpha = 2.280 \text{ rad/s}^2$$

$$a_A = \alpha r_C = (2.280)(1) = 2.280 \text{ ft/s}^2.$$

(c) $\boxed{\vec{a}_A = 2.28 \text{ ft/s}^2 \downarrow}$ 8% less than with a massless pulley.

Example 20.2

In Figure 20.8 Newtdog is on a 2-m-long slender beam with mass 500 kg. We'll ignore Newt-dog's weight. He cuts the rope suspending one end. Determine the tension in rope A just after he cuts rope B.

We'll want to know the mMoI of the beam. Because it is called "slender" we use the mMoI equation for a rod rather than a rectangular prism: $\overline{I}_{AB} = \frac{1}{12}ml^2 = \frac{1}{12}(500)(2)^2 = 166.7 \text{ kg} \cdot \text{m}^2$. If we'd been given the height of the beam, say 150 mm tall, we'd have used

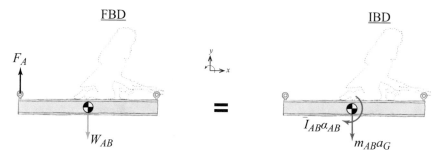

Figure 20.9: FBD/IBD for dynamics of Example 20.2 (© E. Diehl).

$\overline{I}_{AB} = \frac{1}{12}m\left(l^2 + h^2\right) = \frac{1}{12}(500)\left((2)^2 + (0.15)^2\right) = 167.6 \text{ kg} \cdot \text{m}^2$ which is 0.56% more, so assuming it's slender for a beam that tall is a fair estimate.

To evaluate the tension after rope B is cut we draw an FBD/IBD pair. Note that this is "impending motion" which is reminiscent of the exaggeration depicted by a certain cartoon coyote. We can imagine this infinitesimally brief instant in time where the rope is cut but there isn't any motion yet. The next instant there is velocity, therefore there is acceleration:

$$\uparrow \sum F_y = ma_y \quad F_A - W_{AB} = -m_{AB}a_G$$

$$F_A - (500)(9.81) = -(500)a_G \quad \frac{(4{,}905) - F_A}{(500)} = a_G \quad \text{①}$$

$$\circlearrowright \sum M_G = \overline{I}_{AB}\alpha_{AB} \quad -F_A L/2 = -\overline{I}_{AB}\alpha_{AB}.$$

We note that $a_G = L/2\alpha_{AB}$ so $\alpha_{AB} = \frac{a_G}{L/2}$

$$-F_A(2/2) = -(166.7)\frac{a_G}{(2/2)} \quad \frac{F_A}{(166.7)} = a_G \quad \text{②}.$$

Set equations ① and ② equal:

$$\frac{(4{,}905) - F_A}{(500)} = \frac{F_A}{(166.7)} \quad (166.7)(4{,}905) - (166.7)F_A = (500)F_A \quad F_A = 1{,}226 \text{ N}.$$

For comparison to the pre-cut tension, we note:

$$2F_{A,before} - (500)(9.81) = 0, \quad F_{A,before} = 2{,}453 \text{ N}.$$

We conclude that cutting one rope causes the other rope to have LESS tension than beforehand. This is definitely counterintuitive, as we'd likely expect it would have more tension than before

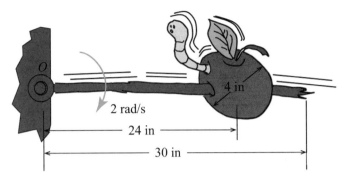

Figure 20.10: Wormy's apple on a stick in Example 20.3 (© E. Diehl).

the cut. The impending motion is rotation acceleration, clockwise, so the rope is being relieved of some of the load it was carrying pre-cut.

Answer: $\boxed{F_A = 1{,}226 \text{ N}}$.

Example 20.3

At the instant shown in Figure 20.10, Wormy's apple is fixed onto a stick rotating in the vertical plane on a frictionless pin with a clockwise angular velocity of $\omega = 2$ rad/s. The apple (and Wormy) are represented by a homogenous 4-in-diameter sphere weighing 8 oz. The stick is 30 in long, weighs one pound, and the apple is 24 in from the pin. Determine the reaction forces at the pin.

There are two ways to approach this problem using N2L: by finding the mass moment of inertia of the stick and apple together about the pin or by parts. The former is simpler so we'll do that first.

The mass of the apple is:

$$m_A = \frac{(8/16)}{(32.2)} = 1.553E - 2 \text{ slugs}.$$

The mass of the stick is:

$$m_S = \frac{(1)}{(32.2)} = 3.106E - 2 \text{ slugs}.$$

The center of mass is found from

$$\bar{x} = \frac{\sum m_i \cdot x_i}{\sum m_i} = \frac{(1.553E - 2)(24/12) + (3.106E - 2)(15/12)}{(1.553E - 2) + (3.106E - 2)} = 1.500 \text{ ft} = r_{G/O}.$$

Table 20.1: Center of mass with respect to pin in Example 20.3

Part	m_i	x_i	$m_i \cdot x_i$
Apple	1.553E - 2	24/12	3.106E - 2
Stick	3.106E - 2	15/12	3.883E - 2
$\sum m_i =$	4.659E - 2	$\sum m_i \cdot x_i =$	6.989E - 2

Table 20.2: Mass moment of inertia about the pivot point in Example 20.3

Part	m_i	L or r	$\overline{I}i$	x_i	$m \cdot x^2$	$\overline{I} + m \cdot x^2$
Apple	1.553E - 2	2/12	$\frac{2}{5}mr^2 = \frac{2}{5}(1.553E\text{-}2)(2/12)^2$ $= 4.314E\text{-}4$	24/12	(1.553E - 2) $\cdot (24/12)^2$ = 6.132E - 2	6.175E - 2
Stick	3.106E - 2	30/12	$\frac{1}{12}mL^2 = \frac{1}{12}(3.106E\text{-}2)$ $(30/12)^2 = 1.618E\text{-}2$	15/12	(3.106E - 2) $\cdot (15/12)^2$ = 4.853E - 2	6.471E - 2
					$I_o =$	1.265E - 1

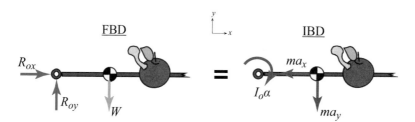

Figure 20.11: FBD/IBD pair of Example 20.3 using mMol about pin (© E. Diehl).

As demonstrated in Section 19.1, a table such as Table 20.1 helps organize finding the center of mass.

The mMoI about the pivot Point O is found using Table 20.2 (note: the mMoI of a sphere is $\overline{I}_{sphere} = \frac{2}{5}mr^2$)

$$I_O = 1.265E - 1 \text{ slug} \cdot \text{ft}^2.$$

We make an FBD/IBD pair to of the apple and stick (Figure 20.11) drawing the reactions as if they are in the positive direction by default:

Rotation:

$$\circlearrowleft \sum M_O = \sum (M_O)_{\text{eff}}$$

$$- mgr_{G/O} = -I_O \alpha$$

$$- (4.659E - 2)(32.2)(1.500) = -(1.265E - 1)\alpha$$

$$\alpha = 17.79 \text{ rad/s}^2 \ \circlearrowleft \ .$$

Translation:

We know the acceleration in the x-direction is the normal acceleration at the center of mass and with this we can find the reaction at the pin in the x-direction:

$$a_x = a_n = \omega^2 r = (2)^2 (1.5) = 6.000 \text{ ft/s}^2 \ \leftarrow$$

$$\rightarrow \sum F_x = ma_x$$

$$R_{Ox} = -(4.659E - 2)(6.000) \quad R_{Ox} = -0.2741 \text{ lb} = 0.2741 \text{ lb} \ \leftarrow \ .$$

The acceleration in the y-direction is the tangential acceleration at the center of mass:

$$a_y = a_t = \alpha r = (17.79)(1.5) = 26.69 \text{ ft/s}^2 \ \downarrow \ .$$

Compare this to the acceleration of gravity (the acceleration it would travel except the mMoI resists this acceleration). With this acceleration we can find the reaction at the pin in the y-direction:

$$\uparrow \sum F_y = ma_y$$

$$R_{Oy} - (4.659E - 2)(32.2) = -(4.659E - 2)(26.69) \qquad R_{Oy} = 0.2567 \text{ lb} \ \uparrow \ .$$

Answers: $\boxed{R_{Ox} = 0.274 \text{ lb} \ \leftarrow}$ $\boxed{R_{Oy} = 0.257 \text{ lb} \ \uparrow}$.

An alternate approach to find the angular acceleration is to take each piece individually (instead of using the center of mass and P.A.T.).

First we establish the kinematic relationships for the individual parts using the center of each of the part. We also find the mMoI of each part:

$$r_{S/O} = 15 \text{ in} = 1.25 \text{ ft} \qquad r_{A/O} = 2 \text{ ft}$$

$$a_{Sy} = (a_S)_t = \alpha r_{S/O} = \alpha (1.25) \qquad a_{Ay} = (a_A)_t = \alpha r_{A/O} = \alpha (2)$$

$$a_{Sx} = (a_S)_n = \omega^2 r_{S/O} = (2)^2 (1.25) = 5.000 \text{ ft/s}^2$$

$$a_{Ax} = (a_A)_t = \omega^2 r_{A/O} = (2)^2 (2) = 8.000 \text{ ft/s}^2$$

$$\bar{I}_S = \frac{1}{12} m_S L^2 = \frac{1}{12} (3.106E - 2)(30/12)^2 = 1.618E - 2 \text{ slug} \cdot \text{ft}^2$$

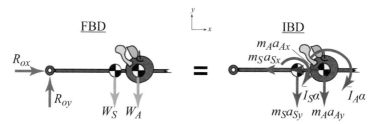

Figure 20.12: FBD/IBD pair of Example 20.3 using mMol about each part and effective inertial moments (© E. Diehl).

$$\bar{I}_A = \frac{2}{5}m_A r^2 = \frac{2}{5}(1.553E-2)(2/12)^2 = 4.314E-4 \text{ slug} \cdot \text{ft}^2.$$

We make another FBD/IBD pair of the apple and stick, this time with individual parts:

$$\circlearrowleft \sum M_O = \sum (M_O)_{eff}$$

$$-m_S \cdot g \cdot r_{S/O} - m_A \cdot g \cdot r_{A/O} = -\bar{I}_S \alpha - m_S a_{Sy} r_{S/O} - \bar{I}_A \alpha - m_A a_{Ay} r_{A/O}$$

$$-(3.106E-2)(32.2)(1.25) - (1.553E-2)(32.2)(2)$$
$$= -(1.618E-2)\alpha - (3.106E-2)\alpha(1.25)(1.25)$$
$$-(4.314E-4)\alpha - (1.553E-2)\alpha(2)(2)$$

$$\alpha = 17.79 \text{ rad/s}^2 \circlearrowleft .$$

We get the same result as with treating the stick and apple together as one mMoI about the pin. You should look carefully at the two approaches and recognize that the P.A.T. is hidden within the effective inertial moment portion of N2L.

An important aspect to using the first approach is to recognize you don't need to account for the effective moment due to the mass times acceleration through the center of mass because it is already addressed when you assembled the mMoI about the pin. But this is only applicable in the special case of rotation about a fixed point. When there is no fixed point, the effective moment concept is needed. We will demonstrate an example of this in Class 21, notably in Example 21.2 where a rod is moving in both translation and rotation.

Example 20.4
The drawbridge in Figure 20.13 is $L = 5$ m long and weighs $m = 1,000$ kg. A single cable is attached to the end and in the position shown $\theta = 45°$. Starting from rest in the position shown, the winch pulls the cable at a constant tension $F_T = 10$ kN. Find (a) the acceleration at Point B (in m/s^2) and (b) the reaction forces at pin O (in Newtons).

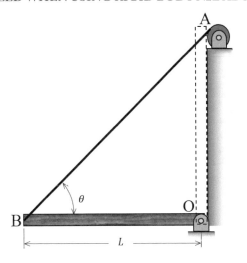

Figure 20.13: Drawbridge in Example 20.4.

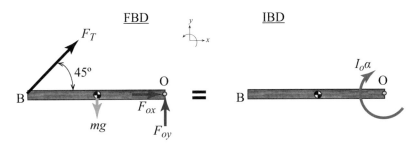

Figure 20.14: FBD/IBD of drawbridge about pin in Example 20.4.

The mMoI of the drawbridge about Point O:

$$\overline{I} = \frac{1}{12}mL^2 = \frac{1}{12}(1{,}000)(5)^2 = 2{,}083 \text{ kg} \cdot \text{m}^2$$

$$I_O = \overline{I} + m\left(\frac{L}{2}\right)^2 = (2{,}083) + (1{,}000)\left(\frac{5}{2}\right)^2 = 8{,}333 \text{ kg} \cdot \text{m}^2.$$

The FBD/IBD pair for this initial position is shown in Figure 20.14:

$$\circlearrowleft \sum M_O = \sum (M_O)_{\text{eff}} = I_O \alpha$$

$$-F_T \sin\theta \cdot L + mg\frac{L}{2} = -I_O \alpha$$

$$-(10{,}000)\sin(45°)(5) + (1{,}000)(9.81)\frac{(5)}{2} = -(8{,}333)\alpha$$

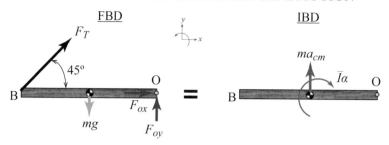

Figure 20.15: FBD/IBD of drawbridge using effective moment in Example 20.4.

$$\alpha = 1.300 \text{ rad/s}^2$$

$$a_B = \alpha L = (1.300)(5) = 6.498 \text{ m/s}^2$$

$$\boxed{\vec{a}_B = 6.50 \text{ m/s}^2 \uparrow}$$

$$\rightarrow \sum F_x = ma_x \quad F_T \cos\theta + F_{Ox} = 0 \quad (10{,}000)\cos\left(45°\right) + F_{Ox} = 0$$

$$\boxed{F_{Ox} = 7{,}071 \text{ N} \leftarrow}$$

$$\uparrow \sum F_y = ma_y \quad F_T \sin\theta - mg + F_{Oy} = ma\frac{L}{2}$$

$$(10{,}000)\sin\left(45°\right) - (1{,}000)(9.81) + F_{Oy} = (1{,}000)(1.300)\frac{(5)}{2}$$

$$\boxed{F_{Oy} = 5{,}990 \text{ N} \uparrow}.$$

Alternatively, we could also have not shifted the mMoI from the center to the pin and instead used the effective moment due to the acceleration at the center of mass, as shown below in the FBD/IBD of Figure 20.15:

$$\circlearrowleft \sum M_O = \sum (M_O)_{\text{eff}}$$

$$- F_T \sin\theta \cdot L + mg\frac{L}{2} = -\bar{I}_O\alpha - ma_{cm}\frac{L}{2}$$

$$a_{cm} = \alpha\frac{L}{2}$$

$$- (10{,}000)\sin\left(45°\right)(5) + (1{,}000)(9.81)\frac{(5)}{2} = -(2{,}083)\,\alpha - (1000)\,\alpha\frac{(5)}{2}\frac{(5)}{2}$$

$$\alpha = 1.300 \text{ rad/s}^2$$

$$a_B = \alpha L = (1.300)(5) = 6.498 \text{ m/s}^2.$$

This returns the same result.

Book 2 - Class 21

https://www.youtube.com/watch?v=TP71bpYVj9g

CLASS 21

Newton's Second Law in Translation and Rotation Plane Motion

B.L.U.F. (Bottom Line Up Front)

- More N2L for Rigid Body Motion Kinetics.

- It's often necessary to break apart an assembly made up of multiple parts and apply N2L to each.

- Rigid Body Kinematics is often a useful tool to solve complex Kinetics problems.

21.1 N2L OF A RIGID BODY IN BOTH TRANSLATION AND ROTATION

In the previous chapter we demonstrated Rigid Body N2L on objects constrained about a fixed axis. Here we apply Rigid Body N2L in more general cases including somewhat more complicated scenarios.

Example 21.1 (related to Examples 14.1, 15.1, and 16.1)
Newtdog is riding a penny-farthing bicycle in Figure 21.1 which has a $d = 4$ ft front wheel. The bicycle is traveling at $v_A = 4$ ft/s, and the front wheel, which is slipping, is rotating at $\omega_{AB} = 3$ rad/s clockwise. The bicycle is also accelerating at $a_A = 1$ ft/s^2 while Newtdog increases his pedaling rate by $\alpha_{AB} = 3$ rad/s^2. Assume the front wheel weighs 50 lb with a 1.75-ft radius of gyration, Newtdog weighs 150 lb with his center of mass 5 ft off the ground, the back wheel is 1 ft diameter and weighs 20 lb with a 0.375-ft radius of gyration, and the remainder of the bicycle weighs 50 lb with a center of gravity 3 ft off the ground. Determine the moment Newtdog needs to apply with the pedals to achieve this acceleration and the power he is generating.

The mMoIs of the wheels are:

$$\overline{I}_{AB} = m_{AB}k_{AB}^2 = \left(\frac{50}{32.2}\right)(1.75)^2 = 4.755 \text{ slug} \cdot \text{ft}^2$$

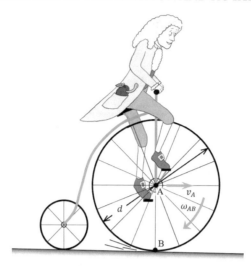

Figure 21.1: Newtdog on a penny-farthing in Example 21.1 (repeat of Figures 14.10, 15.8, and 16.6) (© E. Diehl).

$$\overline{I}_{CD} = m_{CD}k_{CD}^2 = \left(\frac{20}{32.2}\right)(0.375)^2 = 8.734E - 2 \text{ slug} \cdot \text{ft}^2.$$

All of the translational accelerations are the same, so

$$a_A = a_C = a_{PF} = a_{ND} = 1 \text{ ft/s}^2.$$

We note that the rear wheel won't slip so its angular acceleration is

$$\alpha_{CD} = a_A/r_{CD} = (1)/(0.5) = 2 \text{ rad/s}^2.$$

The FBD/IBD pair of the rear wheel is shown in Figure 21.2:

$$\circlearrowleft \sum M_D = \sum (M_D)_{eff}$$

$$- F_{Cx}r_{CD} = -\overline{I}_{CD}\alpha_{CD} - m_{CD}a_C r_{CD}$$

$$- F_{Cx}(0.5) = -(8.734E - 2)(2) - \left(\frac{20}{32.2}\right)(1)(0.5)$$

$$F_{Cx} = 0.9705 \text{ lb} \rightarrow .$$

The FBD/IBD pair of Newtdog and the bicycle without its wheels is shown in Figure 21.3.

$$\rightarrow \sum F_x = ma_x$$

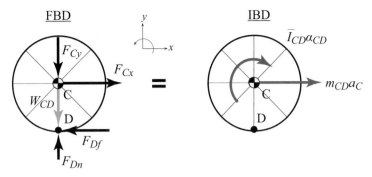

Figure 21.2: FBD/IBD of rear wheel in Example 21.1.

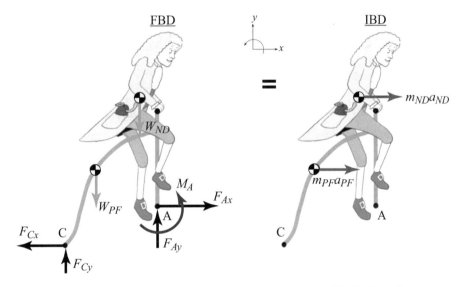

Figure 21.3: FBD/IBD of Newtdog on bicycle in Example 21.1 (© E. Diehl).

$$- F_{Cx} + F_{Ax} = m_{PF}a_{PF} + m_{ND}a_{ND}$$

$$- (0.9705) + F_{Ax} = \left(\frac{50}{32.2}\right)(1) + \left(\frac{150}{32.2}\right)(1)$$

$$F_{Ax} = 7.182 \text{ lb} \rightarrow .$$

The FBD/IBD pair of the front wheel is shown in Figure 21.4.

$$\circlearrowright \sum M_B = \sum (M_B)_{eff}$$

$$F_{Ax}r_{AB} - M_A = -\overline{I}_{AB}\alpha_{AB} - m_{AB}a_A r_{AB}$$

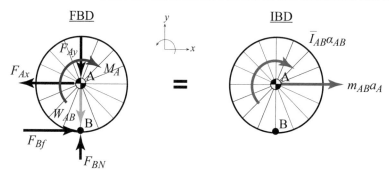

Figure 21.4: FBD/IBD of bicycle front wheel in Example 21.1.

$$(7.182)\,(2) - M_A = -\,(4.755\,)\,(3) - \left(\frac{50}{32.2}\right)(1)\,(2)$$

$$M_A = 31.74\ \text{ft} \cdot \text{lb}.$$

Jumping ahead to Section 22.2 we find the equation for power from torque and rpm. The front wheel is turning at $\omega_{AB} = 3$ rad/s which is $n = (3\ \text{rad/s})\,(60\ \text{s/min})/(2\pi\ \text{rad/rev}) = 28.65$ rpm:

$$\mathbb{P}\,(\text{hp}) = \frac{T\,(\text{ft} \cdot \text{lb})\,n\,(\text{rpm})}{5{,}252} = \frac{(31.74)\,(28.65)}{5{,}252} = 0.1731\ \text{hp}.$$

Answers: $\boxed{M_A = 31.7\ \text{ft} \cdot \text{lb}}$ and $\boxed{\mathbb{P} = 0.173\ \text{hp}}$.

We note that in accordance with Newton's Third Law (N3L!) we needed to make forces equal and opposite on the parts when we separate them. Also, we note that the power is quite small, even compared to a weed trimmer, although this is delivered power.

Example 21.2   (related to Examples 14.2, 15.2, and 16.2)
Link ABC shown in Figure 21.5 has wheels at Points A and B that remain in contact on the horizontal and vertical surfaces. The dimensions are $L_{AB} = 4$ ft and $L_{BC} = 6$ ft and it weighs 100 lb. The link is released from rest when $\theta = 25°$. Assume the wheel masses are negligible and the pins frictionless. Determine the translational and angular accelerations of the rod at this instant.

The mMoI of the link:

$$\overline{I} = \frac{1}{12}mL^2 = \frac{1}{12}\left(\frac{100}{32.2}\right)(10)^2 = 25.88\ \text{slug} \cdot \text{ft}^2.$$

The FBD/IBD pair is shown in Figure 21.6.

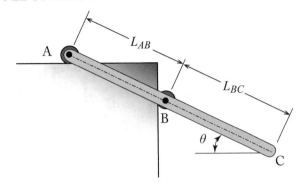

Figure 21.5: Example 21.2 (repeat of Figures 14.12, 15.10, and 16.8).

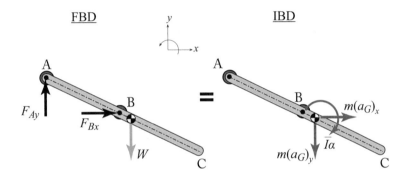

Figure 21.6: FBD/IBD of link in Example 21.2.

$$\circlearrowleft \sum M_A = \sum (M_A)_{eff}$$

$$F_{Bx} L_{AB} \sin\theta - W \frac{(L_{AB} + L_{BC})}{2} \cos\theta$$

$$= -\bar{I}\alpha + m(a_G)_x L_{AB} \sin\theta - m(a_G)_y \frac{(L_{AB} + L_{BC})}{2} \cos\theta$$

$$F_{Bx}(4)\sin(25°) - (100)\frac{((4)+(6))}{2}\cos(25°)$$

$$= -(25.88)\alpha + \left(\frac{100}{32.2}\right)(a_G)_x (4)\sin(25°) - \left(\frac{100}{32.2}\right)(a_G)_y \frac{((4)+(6))}{2}\cos(25°)$$

$$F_{Bx}(1.690) - (453.2) = -(25.88)\alpha + (5.250)(a_G)_x - (14.07)(a_G)_y. \quad \text{①}$$

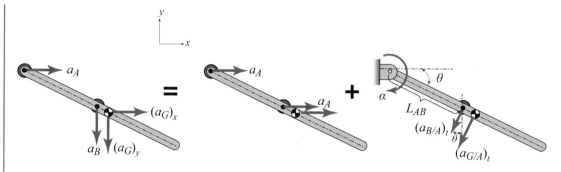

Figure 21.7: Acceleration diagram of Example 21.2.

So far, there are four unknowns. We can also find the balance of forces in the x-direction. The y-direction only adds another unknown, so we won't pursue that.

$$\rightarrow \sum F_x = ma_x$$

$$F_{Bx} = m(a_G)_x$$

$$F_{Bx} = \left(\frac{100}{32.2}\right)(a_G)_x$$

$$F_{Bx} = (3.106)(a_G)_x. \quad ②$$

We can use kinematics to relate the accelerations. Figure 21.7 is the acceleration diagram of the link.

Setting up the relation of acceleration of Point B using A as the reference:

$$\vec{a}_B = \vec{a}_A + (\vec{a}_{B/A})_t + (\vec{a}_{B/A})_n .$$

x-dir:

$$(a_B)_x = (a_A)_x + ((a_{B/A})_t)_x + ((a_{B/A})_n)_x$$

$$(0) = a_A - \alpha L_{AB} \sin\theta + (0)$$

$$a_A = \alpha (4) \sin (25°)$$

$$a_A = (1.690)\alpha. \quad ③$$

Now we set up the acceleration of the center of gravity.

$$\vec{a}_G = \vec{a}_A + (\vec{a}_{G/A})_t + (\vec{a}_{G/A})_n .$$

x-dir:

$$(a_G)_x = (a_A)_x + ((a_{G/A})_t)_x + ((a_{G/A})_n)_x$$

$$(a_G)_x = a_A - \alpha \frac{(L_{AB} + L_{BC})}{2} \sin \theta + (0)$$

$$(a_G)_x = (1.690) \alpha - \alpha \frac{((4) + (6))}{2} \sin (25°)$$

$$(a_G)_x = - (0.4226) \alpha. \quad \text{④}$$

y-dir:

$$(a_G)_y = (a_A)_y + ((a_{G/A})_t)_y + ((a_{G/A})_n)_y$$

$$- (a_G)_y = (0) - \alpha \frac{(L_{AB} + L_{BC})}{2} \cos \theta + (0)$$

$$(a_G)_y = \alpha \frac{((4) + (6))}{2} \cos (25°)$$

$$(a_G)_y = (4.532) \alpha. \quad \text{⑤}$$

We can combine equations ①, ②, ④, and ⑤ to find the angular acceleration:

$$(3.106) (- (0.4226)) \alpha (1.690) - (453.2)$$

$$= - (25.88) \alpha + (5.250) (- (0.4226)) \alpha - (14.07) (4.532) \alpha$$

$$\alpha = 5.055 \text{ rad/s}^2 \ \circlearrowleft$$

$$(a_G)_x = - (0.4226) (5.055) = 2.136 \text{ ft/s}^2 \ \leftarrow$$

$$(a_G)_y = (4.532) (5.055) = 22.91 \text{ ft/s}^2 \ \downarrow$$

$$|a_G| = \sqrt{(2.136)^2 + (22.91)^2} = 23.01 \text{ ft/s}^2$$

$$\theta = \tan^{-1} \left(\frac{22.91}{2.136} \right) = 84.67°.$$

Answers: $\boxed{\vec{a}_G = 23.0 \text{ ft/s}^2 \ 84.7° \ \nearrow}$ $\boxed{\alpha = 5.06 \text{ rad/s}^2 \ \circlearrowleft}$.

In Example 16.2 we were given velocity and an acceleration (actually a deceleration) without knowing why it was occurring. This example is different firstly because it starts from rest, and secondly because we now know the reason it is moving: gravity. In Example 16.2, there could have been some external load on the rod, but all we were doing was describing the motion: the difference between kinematics and kinetics.

Example 21.3   (related to Examples 14.3, 15.3, and 16.3)

Crank BC of the slider-crank in Figure 21.8 rotates at a constant $\omega_{BC} = 8$ rad/s clockwise and is at $\theta = 40°$. The dimensions are $L_{AB} = 500$ mm, $L_{BC} = 300$ mm, and $L_{BD} = 150$ mm. A force

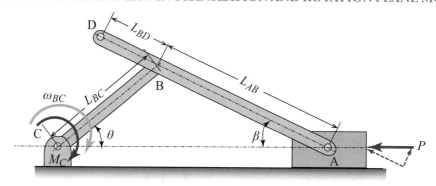

Figure 21.8: Example 16.3 (repeat of Figures 14.16 and 15.12).

FBD   FBD

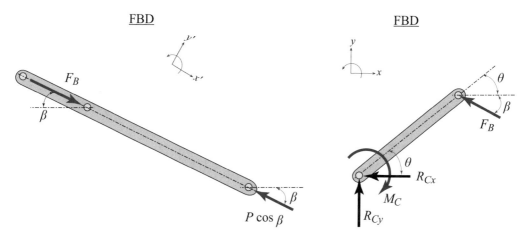

Figure 21.9: Static FBDs of Example 21.3.

due to pressure on the face of piston A is 100 N. Rod ABD has a mass of $m_{ABD} = 6$ kg. Treat the mass of crank BC and piston A as negligible. Determine the torque on pin C due to statics and dynamics.

We found $\beta = 22.69°$ and $\omega_{ABD} = 3.985$ rad/s ↺ in Example 14.3. We found $\vec{a}_B = 1.920$ m/s^2 40° ↗, and $\alpha_{ABD} = 20.11$ rad/s^2 ↻ in Example 16.3. If we hadn't already found these in the previous examples we would do that first.

Figure 21.9 presents the statics FBDs. The connecting rod ABD is a 2-force member since both connections are pins that do not support a moment. The FBD of the crank BC is not a 2-force member since the connection at C carries the input moment (torque). Figure 21.9 presents the statics FBDs.

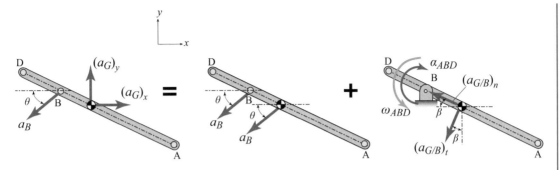

Figure 21.10: Acceleration diagram of connecting rod of Example 21.3.

$$\searrow \sum F_{x'} = 0$$

$$F_B - P\cos\beta = 0$$

$$F_B - (100)\cos(22.69°) = 0$$

$$F_B = 92.26 \text{ N}$$

We want to avoid the reaction forces at C so we take the summation of moments about it:

$$\circlearrowleft \sum M_C = 0$$

$$-M_C + F_B\sin(\theta + \beta)L_{BC} = 0$$

$$-M_C + (92.26)\sin((40°) + (22.69°))(0.300) = 0$$

$$M_C = 2.459 \text{ Nm} \circlearrowleft.$$

To find the force at B for the dynamics case we first need to find the acceleration of the center of mass of connecting rod ABD using the acceleration diagram in Figure 21.10. Note that it is NOT a 2-force member because the effective inertial forces must also be considered (since there are more than just two forces). The components of the acceleration at G are drawn in the positive direction by default:

$$\vec{a}_G = \vec{a}_B + \left(\vec{a}_{G/B}\right)_t + \left(\vec{a}_{G/B}\right)_n.$$

x-dir:

$$(a_G)_x = (a_B)_x + \left((a_{G/B})_t\right)_x + \left((a_{G/B})_n\right)_x$$

$$(a_G)_x = -a_B\cos\theta - \alpha_{ABD}r_{G/B}\sin\beta - \omega_{ABD}^2 r_{G/B}\cos\beta$$

$$(a_G)_x = -(1.920)\cos(40°) - (20.11)(0.175)\sin(22.69°) - (3.985)^2(0.175)\cos(22.69°)$$

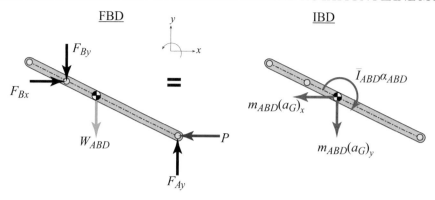

Figure 21.11: FBD/IBD of connecting rod in Example 21.3.

$$(a_G)_x = -5.392 \text{ m/s}^2.$$

y-dir:

$$(a_G)_y = (a_B)_y + \left(\left(a_{G/B}\right)_t\right)_y + \left(\left(a_{G/B}\right)_n\right)_y$$

$$(a_G)_y = -a_B \sin\theta - \alpha_{ABD}r_{G/B}\cos\beta + \omega_{ABD}^2 r_{G/B}\sin\beta$$

$$(a_G)_y = -(1.920)\sin\left(40°\right) - (20.11)(0.175)\cos\left(22.69°\right) + (3.985)^2(0.175)\sin\left(22.69°\right)$$

$$(a_G)_y = -3.409 \text{ m/s}^2.$$

With the kinematics out of the way, we can address the kinetics. The mMoI of the connecting rod is

$$\bar{I}_{ABD} = \frac{1}{12}m_{ABD}L_{ABD}^2 = \frac{1}{12}(6)(0.650)^2 = 0.2113 \text{ kg}\cdot\text{m}^2.$$

The FBD/IBD pair of the connecting rod is shown in Figure 21.11.

$$\rightarrow \sum F_x = ma_x$$

$$F_{Bx} - P = m_{ABD}(a_G)_x$$

$$F_{Bx} - (100) = -(6)(5.392)$$

$$F_{Bx} = 67.65 \text{ N} \rightarrow$$

We take the summation of moments about A so we can avoid finding the vertical reaction force there. Remember we must use the "effective inertial moments" and include the mass times acceleration at the center times distance to A:

$$\circlearrowleft \sum M_A = \sum (M_A)_{eff}$$

FBD

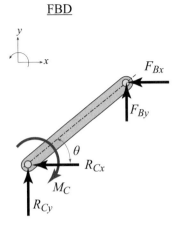

Figure 21.12: FBD of crank BC in Example 21.3.

$$- F_{Bx} L_{AB} \sin \beta - F_{By} L_{AB} \cos \beta + m_{ABD} g r_{G/A} \cos \beta$$
$$= -\overline{I}_{ABD} \alpha_{ABD} + m_{ABD} (a_G)_x r_{G/A} \sin \beta + m_{ABD} (a_G)_y r_{G/A} \cos \beta$$

$$- (67.65)(0.500) \sin(22.69°) + F_{By}(0.500) \cos(22.69°) + (6)(9.81)(0.325) \cos(22.69°)$$
$$= -(0.2113)(20.11) + (6)(5.392)(0.325) \sin(22.69°) + (6)(3.409)(0.325) \cos(22.69°)$$

$$F_{By} = 22.12 \text{ N} \downarrow .$$

Because crank BC is treated as massless, we can use an FBD rather than an FBD/IBD pair in Figure 21.12 to evaluate the loads on it. Remember that the forces at B are equal and opposite from the previous FBD:

$$\circlearrowright \sum M_C = 0$$
$$- M_C + F_{Bx} L_{BC} \sin(\theta) + F_{By} L_{BC} \cos(\theta) = 0$$
$$- M_C + (67.65)(0.300) \sin\left(40°\right) + (22.12)(0.300) \cos\left(40°\right) = 0$$
$$M_C = 18.13 \text{ Nm} \circlearrowright .$$

Answers:

Torque due to statics $\boxed{M_C = 2.46 \text{ Nm} \circlearrowright}$.

Torque due to dynamics $\boxed{M_C = 18.1 \text{ Nm} \circlearrowright}$.

We conclude that it takes quite a bit more moment (torque) to turn the crank dynamically even when the load at P is the same and the crank is at constant velocity.

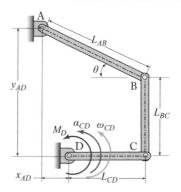

Figure 21.13: Example 21.4.

Example 21.4 (related to Examples 14.4, 15.4, and 16.4)
In the position shown, bar CD of the four-bar mechanism in Figure 21.13 has an angular velocity of $\omega_{CD} = 12$ rad/s counter-clockwise and accelerating at $\alpha_{CD} = 25$ rad/s^2 counter-clockwise. The dimensions are $L_{AB} = 600$ mm, $L_{BC} = 400$ mm, $L_{CD} = 350$ mm, and $x_{AD} = 200$ mm. Each bar is 40 mm wide, 10 mm thick, and is made of steel (7700 kg/m^3). The mechanism operates in the horizontal plane ($\because$ gravitational forces aren't present). Determine the torque applied at pin D to cause this motion.

We first do some house keeping and find the masses and mMoIs. Mass properties of links:

$$m_{AB} = \rho L_{AB} wt = (7700)(0.6)(0.040)(0.01) = 1.848 \text{ kg}$$

$$\overline{I}_{AB} = \frac{1}{12} m_{AB} L_{AB}^2 = \frac{1}{12}(1.848)(0.6)^2 = 0.05544 \text{ kg} \cdot \text{m}^2$$

$$I_A = \overline{I}_{AB} + m_{AB}(L_{AB}/2)^2 = (0.05544) + (1.848)(0.3)^2 = 0.2218 \text{ kg} \cdot \text{m}^2$$

$$m_{BC} = \rho L_{BC} wt = (7700)(0.4)(0.040)(0.01) = 1.232 \text{ kg}$$

$$\overline{I}_{BC} = \frac{1}{12} m_{BC} L_{BC}^2 = \frac{1}{12}(1.232)(0.4)^2 = 0.01643 \text{ kg} \cdot \text{m}^2$$

$$m_{CD} = \rho L_{CD} wt = (7700)(0.35)(0.040)(0.01) = 1.078 \text{ kg}$$

$$\overline{I}_{CD} = \frac{1}{12} m_{CD} L_{CD}^2 = \frac{1}{12}(1.078)(0.35)^2 = 0.01101 \text{ kg} \cdot \text{m}^2$$

$$I_C = \overline{I}_{CD} + m_{CD}(L_{CD}/2)^2 = (0.01101) + (1.078)(0.35)^2 = 0.04402 \text{ kg} \cdot \text{m}^2.$$

From Example 14.4:

$$\omega_{AB} = 7.637 \text{ rad/s} \circlearrowleft, \quad \omega_{BC} = 4.579 \text{ rad/s} \circlearrowright, \quad \text{and} \quad \theta = 23.56°.$$

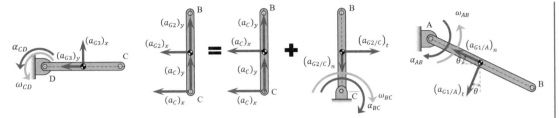

Figure 21.14: Acceleration diagrams of (a) crank CD, (b) coupler BC, and rocker AB of Example 21.4.

From Example 16.4 (confirmed in Example 17.3):

$$\alpha_{AB} = 24.77 \text{ rad/s}^2 \circlearrowleft, \quad \alpha_{BC} = 30.99 \text{ rad/s}^2 \circlearrowleft,$$
$$(a_C)_x = 50.40 \text{ m/s}^2 \leftarrow, \quad \text{and} \quad (a_C)_y = 8.750 \text{ m/s}^2 \uparrow .$$

We need the acceleration at the mass center of each link using kinematics. The acceleration diagrams for the three links are shown in Figure 21.14. The centers are labeled as $G1$ for rocker AB, $G2$ for coupler BC, and $G3$ for crank CD.

For crank CD in Figure 21.14a:

$$(a_{G3})_x = (a_{G3/D})_n = \omega_{CD}^2 L_{CD}/2 = (12)^2(0.350)/2 = 25.20 \text{ m/s}^2 \leftarrow$$

$$(a_{G3})_y = (a_{G3/D})_t = \alpha_{CD} L_{CD}/2 = (25)(0.350)/2 = 4.375 \text{ m/s}^2 \uparrow .$$

For coupler BC in Figure 21.14b:

$$\vec{a}_{G2} = \vec{a}_C + \left(\vec{a}_{G2/C}\right)_t + \left(\vec{a}_{G2/C}\right)_n .$$

x-dir:

$$(a_{G2})_x = (a_C)_x + \left((a_{G2/C})_t\right)_x + \left((a_{G2/C})_n\right)_x$$
$$- (a_{G2})_x = -(a_C)_x + \alpha_{BC} L_{BC}/2 + (0)$$
$$- (a_{G2})_x = -(50.40) + (30.99)(0.400)/2$$
$$(a_{G2})_x = 44.20 \text{ m/s}^2 \leftarrow .$$

y-dir:

$$(a_{G2})_y = (a_C)_y + \left((a_{G2/C})_t\right)_y + \left((a_{G2/C})_n\right)_y$$
$$(a_{G2})_y = (a_C)_y + (0) - \omega_{BC}^2 L_{BC}/2$$
$$(a_{G2})_y = (8.750) + (0) - (4.579)^2(0.400)/2$$

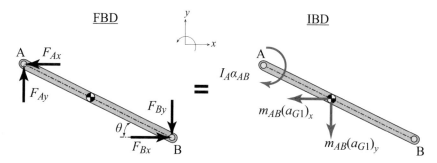

Figure 21.15: FBD/IBD of rocker AB in Example 21.4.

$$(a_{G2})_y = 4.557 \text{ m/s}^2 \uparrow .$$

For rocker AB in Figure 21.14c:

$$(a_{G1})_n = (a_{G1/A})_n = \omega_{AB}^2 L_{AB}/2 = (7.637)^2 (0.6)/2 = 1.750 \text{ m/s}^2 \nwarrow$$

$$(a_{G1})_t = (a_{G1/A})_t = \alpha_{AB} L_{AB}/2 = (24.77)(0.6)/2 = 7.431 \text{ m/s}^2 \swarrow .$$

Resolving to the x and y coordinates (coordinate transformation not shown):

$$(a_{G1})_x = -(a_{G1})_n \cos \theta - (a_{G1})_t \sin \theta$$
$$= -(1.750) \cos (23.56°) - (7.431) \sin (23.56°) = -4.574 \text{ m/s}^2$$

$$(a_{G1})_y = (a_{G1})_n \sin \theta - (a_{G1})_t \cos \theta$$
$$= (1.750) \sin (23.56°) - (7.431) \cos (23.56°) = -6.112 \text{ m/s}^2.$$

With the kinematics complete, the forces can now be found, starting with an FBD/IBD of Rocker AB in Figure 21.15.

Because Rocker AB is in rigid body motion about a fixed Point (A) we use the mMoI about A times the angular acceleration rather than using the effective moments:

$$\circlearrowleft \sum M_A = I_A \alpha_{AB}$$

$$- F_{Bx} L_{AB} \sin \theta - F_{By} L_{AB} \cos \theta = -I_A \alpha_{AB}$$

$$F_{Bx} (0.6) \sin (23.56°) - F_{By} (0.6) \cos (23.56°) = -(0.2218)(24.77)$$

$$(0.2398) F_{Bx} - (0.5500) F_{By} = -(5.494) . \quad ①$$

The FBD/IBD of Coupler BC is shown in Figure 21.16.

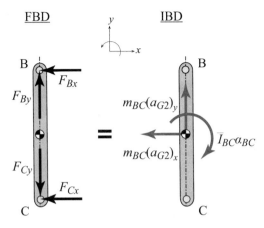

Figure 21.16: FBD/IBD of coupler BC in Example 21.4.

We can find F_{Bx} if we take the moments about Point C. Unlike Rocker AB, we must use the effective moments on Coupler BC since it is not rotating about a fixed point:

$$\circlearrowleft \sum M_C = \sum (M_C)_{eff}$$

$$F_{Bx} L_{BC} = -\overline{I}_{BC}\alpha_{BC} + m_{BC}(a_{G2})_x L_{BC}/2$$

$$F_{Bx}(0.4) = -(0.01643)(30.99) + (1.232)(44.20)(0.4)/2$$

$$F_{Bx} = 10.38 \text{ N}.$$

With this we can find the vertical force at B from equation $\textcircled{1}$:

$$(0.2398)(10.38) - (0.5500) F_{By} = -(5.494)$$

$$F_{By} = 14.52 \text{ N}.$$

We can find the forces at C from the translational N2L of the FBD/IBD of Figure 21.16.

$$\rightarrow \sum F_x = ma_x$$

$$-F_{Cx} - F_{Bx} = -m_{BC}(a_{G2})_x$$

$$-F_{Cx} - (10.38) = -(1.232)(44.20)$$

$$F_{Cx} = 44.07 \text{ N}$$

$$\uparrow \sum F_y = ma_y$$

$$F_{Cy} + F_{By} = m_{BC}(a_{G2})_y$$

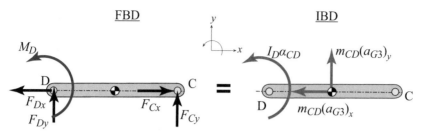

Figure 21.17: FBD/IBD of crank CD in Example 21.4.

$$-F_{Cy} + (14.52) = (1.232)(4.557)$$

$$F_{Cy} = 8.901 \text{ N}.$$

The FBD/IBD of Crank CD is shown in Figure 21.17. We can use rotational N2L about a fixed point to find the required moment (torque) applied at pin D. Note that the vertical force at C is acting in the counter-clockwise direction which will tend to increase the angular acceleration in the counter-clockwise direction:

$$\circlearrowleft \sum M_D = I_D \alpha_{CD}$$

$$M_D + F_C L_{AB} = I_D \alpha_{CD}$$

$$M_D + (8.901)(0.35) = (0.04402)(25)$$

$$M_D = -2.015 \text{ Nm}.$$

The negative sign indicates the moment at D is actually clockwise. How is this possible? This is because Rocker AB is decelerating in this position and requires a load to slow it down which is greater than the inertial forces resulting from the motion of Coupler BC and Rocker CD, therefore a moment at D in the clockwise direction is necessary. The example illustrates the complexity and inter-dependency of rigid body kinetics and kinematics. Generally, a problem any more complicated than this would be approached with other Dynamics techniques. Applying Newtons Laws to challenging problems such as this tests our problem-solving skills and attention to detail.

Answer: $\boxed{M_D = 2.02 \text{ Nm } \circlearrowleft}$.

Book 2 - Class 22

https://www.youtube.com/watch?v=XPSb08hhoQQ

CLASS 22

Energy Methods

B.L.U.F. (Bottom Line Up Front)

- Just as with all rigid body kinetics we include the rotational contribution when applying Work-Energy.

- Kinetic Energy of rigid bodies also includes $KE = \frac{1}{2}\overline{I}\omega^2$.

- Work in rigid bodies also includes the moment integrated against the change in angle $U_{1\rightarrow2} = \int_{\theta_1}^{\theta_2} M d\theta$.

22.1 WORK AND ENERGY APPLIED TO A RIGID BODY

The same principles discussed in Classes 8 (vol. 1) and 9 (vol. 1) for particles can also be applied to Rigid Bodies. Therefore, the energy accounting equation below remains true:

$$\boxed{KE_1 + PE_1 + U_{1\rightarrow2} = KE_2 + PE_2}.$$

Now we must also consider rigid body kinetic energy due to rotational speed, potential energy due to torsion springs and work due to externally applied moments.

The kinetic energy for translation is specific to the center of mass and the rotational term relates mMoI and angular velocity in an analogous way. The equation for kinetic energy in a rigid body is now:

$$\boxed{KE = \frac{1}{2}mv_{CM}^2 + \frac{1}{2}I_{CM}\omega^2}.$$

Just as we had translational work with force through a position change, now we have rotational work with moment through an angular change. The equations for work now become both:

$$\boxed{U_{1\rightarrow2} = \int_{A_1}^{A_2} \overline{\mathbf{F}} \cdot d\overline{\mathbf{r}} = F_{constant}\Delta x} \quad \text{and} \quad \boxed{U_{1\rightarrow2} = \int_{\theta_1}^{\theta_2} M d\theta = M_{constant}\Delta\theta}.$$

This moment is often a torque since it's commonly about a long axis of a shaft or pin. We can easily see in Figure 22.1 that when moment is made up of a force times a radius, then the arc

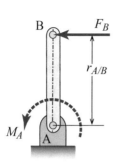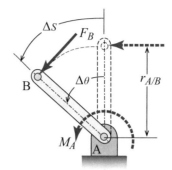

Figure 22.1: Comparion of work from forces and moments.

that it travels is the radius times the angular displacement. The force in this scenario follows the path, so it too is force multiplied by the distance it acts through. Therefore, moment times angular displacement is force times the arc length it travels through which is the definition work.

For potential energy there is no rotational equivalent for gravitational potential energy ($PE_g = Wy = mgh$) but rotational springs (such as in a traditional mouse trap or in a wind-up clock or toy) where the spring constant k_θ has units of $\text{N} \cdot \text{m/rad}$ or $\text{ft} \cdot \text{lb/rad}$:

$$PE_{spr} = \frac{1}{2}kx^2 \quad \text{and} \quad PE_{spr} = \frac{1}{2}k_\theta\theta^2 .$$

For a system of rigid bodies we are able to account for all the energy and external work to evaluate the kinetics:

$$\sum (KE_1)_i + \sum (PE_1)_i + \sum (U_{1\rightarrow2})_i = \sum (KE_2)_i + \sum (PE_2)_i .$$

This can be applied to individual objects *or to* an entire system; a fact which can be used as a strategy to eliminate or ignore the internal forces when treated as a system. As with particles, if there is no external work, $U_{1\rightarrow2} = 0$, then Energy before = Energy after which is the Conservation of Energy:

$$KE_1 + PE_1 = KE_2 + PE_2 .$$

We should always think about this as an energy balance and recognize the trade off between potential energy and kinetic energy. The primary difference in rigid bodies from particles is the inclusion of rotational inertia, as shown in Figure 22.2. The rotational inertia of the drum requires additional energy that must increase as Newtdog's potential energy decreases.

Some caveats to remember when using Work and Energy on rigid bodies are as follows.

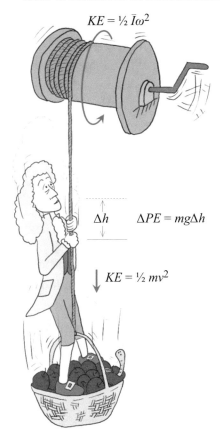

$$KE = \tfrac{1}{2}\,\bar{I}\omega^2$$

$$\Delta h \qquad \Delta PE = mg\Delta h$$

$$KE = \tfrac{1}{2}\,mv^2$$

Figure 22.2: Newtdogs descent slowed by Drum's mMoI (© E. Diehl).

1. The sign convention for work can be established from:

    (a) External work energy input is positive ($+$), work energy removed is negative ($-$).

    (b) Direction of force and displacement agreement.

        i. If the force and distance traveled are in the same direction, work is positive ($+$).

        ii. If the moment couple and angular displacement are in the same direction, work is positive ($+$).

2. The kinetic energy is made up of two parts: *translational* (wrt *CM*) and *rotational* (wrt *CM*)

$$KE = \frac{1}{2}mv_{CM}^2 + \frac{1}{2}I_{CM}\omega^2.$$

3. A special case of (2) is rotation about a fixed point... if the moment of inertia is taken about the fixed Point O, instead of the center of mass, the entire kinetic energy can be

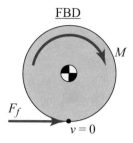

Figure 22.3: FBD of a non-slipping wheel.

reduced to a single equation:

$$KE = \frac{1}{2}I_O\omega^2.$$

4. Work done by the friction of a non-slipping wheel (Figure 22.3) can be ignored because we can rewrite the work equation as a velocity relationship and then consider the velocity at the contact point:

$$U_{1\rightarrow2} = \int \overline{\mathbf{F}} \cdot d\overline{\mathbf{r}} = \int \overline{\mathbf{F}} \cdot \overline{\mathbf{v}}dt.$$

We can also look at the last point as integrating power over time. Since we mention power, we should discuss it with regards to rigid bodies.

22.2   POWER IN RIGID BODIES

As discussed in Section 9.3 (vol. 1) for particles, power is the time rate change of work. We can write the definition of power in a number of ways. Recall from Class 9 (vol. 1) for translation we defined power as:

$$\boxed{\mathbb{P} = \frac{\Delta Work}{\Delta time} = \frac{dU}{dt} = \overline{\mathbf{F}} \cdot \overline{\mathbf{v}}}.$$

For problems involving rotational motion, another form of the power equation is:

$$\boxed{Power = \frac{dU}{dt} = \frac{Md\theta}{dt} = M\omega}.$$

A useful relation for power in machines is often stated as torque multiplied by rpm. Accounting for conversion factors (such as $550\frac{\text{ft·lb/s}}{horsepower}$ and $\frac{(2\pi \text{ rad/rev})}{(60 \text{ s/min})}$) we can write the following for U.S. and S.I. units:

$$\mathbb{P}(\text{hp}) = \frac{T\,(\text{ft} \cdot \text{lb})\,n\,(\text{rpm})}{5{,}252} \qquad \mathbb{P}\,(W) = \frac{T\,(\text{N} \cdot \text{m})\,n\,(\text{rpm})}{9.549}.$$

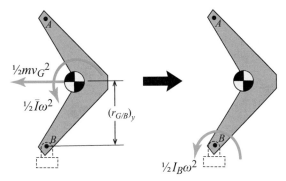

Figure 22.4: Kinetic energy of a boomerang rotating and translating revealing the parallel axis theorem.

22.3 P.A.T. REVEALED WHEN TAKING KINETIC ENERGY ABOUT A FIXED POINT

Similar to what we demonstrated in Section 20.3, the parallel axis theorem can be revealed when applying rigid body kinetic energy to a body rotating about a fixed point. We have the choice to either find the kinetic energy as a combination of rotational energy about the center of mass plus the translational energy of the center of mass, or as the total rotational kinetic energy about the pin. This choice is depicted in Figure 22.4 using our boomerang.

The kinetic energy can be written either way:

$$\frac{1}{2}\overline{I}\omega^2 + \frac{1}{2}mv_G^2 = \frac{1}{2}I_B\omega^2.$$

We note that the velocity of the center of gravity is related to the angular velocity when rotating about a fixed point by $v_G = \omega\left(r_{G/B}\right)_y$ which when substituted:

$$\frac{1}{2}\overline{I}\omega^2 + \frac{1}{2}m\left[\omega\left(r_{G/B}\right)_y\right]^2 = \frac{1}{2}I_B\omega^2$$

$$\frac{1}{2}\left[\overline{I} + m\left(r_{G/B}\right)_y^2\right]\omega^2 = \frac{1}{2}I_B\omega^2.$$

We can see that the P.A.T. is revealed:

$$I_B = \overline{I} + m\left(r_{G/B}\right)_y^2.$$

We have a choice to either use the kinetic energy about the center of mass (which will have two terms: rotation and translation) or use the total rotational kinetic energy about the pin, but we shouldn't mix these up.

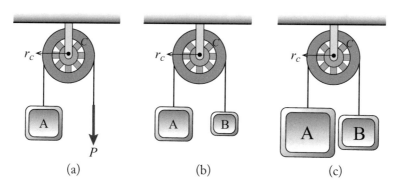

Figure 22.5: Example 22.1 mass and pulley problem (repeat of Figure 20.4).

In Section 8.6 (vol. 1) we outlined a procedure to follow for applying Work-Energy to problems. That same procedure can be followed for rigid bodies as well, but we won't call out the steps as in Class 8 (vol. 1) since we should be comfortable enough with the process to not rely on a step-by-step procedure.

Example 22.1   (This is a repeat of Example 20.1)
Just like Examples 5.3, 8.2, 10.2, (vol. 1), and 20.1, the three weights and pulley setups shown in Figure 22.5 begin at rest. The single frictionless pulley weighs $W_C = 50$ lb, has radius $r_C = 1$ ft and a radius of gyration of $k_C = 0.75$ ft. In setup (a) the force ($P = 50$ lb) is applied to a cable attached to block A ($W_A = 75$ lb). Setup (b) has the same block A and is connected via a cable over the pulley to block B ($W_B = 50$ lb). Setup (c) has larger blocks with the same difference between them ($W_A = 175$ lb and $W_B = 150$ lb). Determine the acceleration of block A for each setup.

The mMoI of the pulley was found in Example 20.1 to be: $\overline{I}_C = 0.8734$ slug $\cdot$ ft^2. Just as with Example 8.2 (vol. 1), this problem asks for acceleration but Work-Energy deals with velocity and displacements so we'll make an assumption about a displacement ($\Delta y = 1$ ft), determine the velocity after it moves that far and then use kinematics to determine the acceleration. We also can use kinematics to relate the translational velocity of the blocks to the angular velocity of the pulley by $v_A = \omega_C r_C$ and $v_B = \omega_C r_C$; therefore, $\omega_C = v_A / r_C$ and $v_A = v_B$.

Part (a) is a standard work-energy problem (not a conservation of energy problem) and the datum (the top of block A in position 2) and displacement are shown in Figure 22.6:

$$KE_1 = 0$$

$$PE_1 = m_A g y_{A1} = W_A y_{A1} = (75)(1) = 75 \text{ ft} \cdot \text{lb}$$

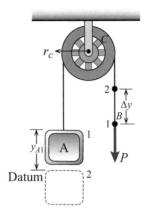

Figure 22.6: Work-energy diagram of Example 22.1(a).

$$U_{1\to 2} = -P\Delta y = -(50)(1) = -50 \text{ ft} \cdot \text{lb} \quad \text{(negative because force and displacement}$$
are in opposite directions)

$$KE_2 = \frac{1}{2}m_A v_{A2}^2 + \frac{1}{2}\bar{I}_C \omega_C^2 = \frac{1}{2}(75/32.2)\,v_{A2}^2 + \frac{1}{2}(0.8734)(v_{A2}/(1))^2 = (1.601)\,v_{A2}^2$$

$$PE_2 = 0$$

$$KE_1 + PE_1 + U_{1\to 2} = KE_2 + PE_2$$

$$(0) + (75) - (50) = (1.601)\,v_{A2}^2 + (0)$$

$$v_{A2} = 3.951 \text{ ft/s}.$$

Using kinematics we can find the constant acceleration (which we know to be constant because all of the forces are constant):

$$v^2 = (v_0)^2 + 2a\Delta y$$

$$(3.951)^2 = (0)^2 + 2a(1)$$

$$a_A = 7.806 \text{ ft/s}^2.$$

(a) $\boxed{\vec{a}_A = 7.81 \text{ ft/s}^2 \downarrow}$ as expected, this matches the results from Example 20.1(a).

Part (b): The second scenario Work-Energy diagram is shown in Figure 22.7 using two different datums (one for each block):

$$KE_1 = 0$$

$$PE_1 = m_A g y_{A1} = W_A y_{A1} = (75)(1) = 75 \text{ ft} \cdot \text{lb}$$

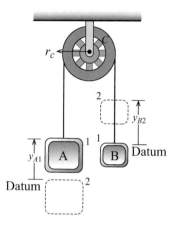

Figure 22.7: Work-energy diagram of Example 22.1(b).

$U_{1\to2} = 0$ (therefore this is a conservation of energy problem)

$$KE_2 = \frac{1}{2}m_A v_{A2}^2 + \frac{1}{2}\bar{I}_C \omega_C^2 + \frac{1}{2}m_B v_{B2}^2$$

$$= \frac{1}{2}(75/32.2)\,v_{A2}^2 + \frac{1}{2}(0.8734)\,(v_{A2}/(1))^2 + \frac{1}{2}(50/32.2)\,v_{A2}^2 = (2.378)\,v_{A2}^2.$$

Note: from the simple pulley kinematics we know that $v_{B2} = v_{A2}$

$$PE_2 = m_B g y_{B2} = W_B y_{B2} = (50)(1) = 50 \text{ ft} \cdot \text{lb}$$

$$KE_1 + PE_1 + U_{1\to2} = KE_2 + PE_2$$

$$(0) + (75) + (0) = (2.378)\,v_{A2}^2 + (50)$$

$$v_{A2} = 3.243 \text{ ft/s}.$$

Using kinematics we can find the constant acceleration (which we know to be constant because all of the forces are constant):

$$v^2 = (v_0)^2 + 2a\,\Delta y$$

$$(3.243)^2 = (0)^2 + 2a\,(1)$$

$$a_A = 5.257 \text{ ft/s}^2.$$

(b) $\boxed{\vec{a}_A = 5.26 \text{ ft/s}^2 \downarrow}$ as expected, this matches the results from Example 20.1(b).

Part (c): The third scenario Work-Energy diagram is shown in Figure 22.8:

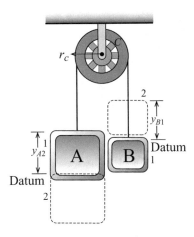

Figure 22.8: Work-energy diagram of Example 22.1(c).

$$KE_1 = 0$$

$$PE_1 = m_A g y_{A1} = W_A y_{A1} = (175)(1) = 175 \text{ ft} \cdot \text{lb}$$

$U_{1\to2} = 0$ (this is also a conservation of energy problem)

$$KE_2 = \frac{1}{2} m_A v_{A2}^2 + \frac{1}{2} \bar{I}_C \omega_C^2 + \frac{1}{2} m_B v_{B2}^2$$

$$= \frac{1}{2}(175/32.2) v_{A2}^2 + \frac{1}{2}(0.8734)(v_{A2}/(1))^2 + \frac{1}{2}(150/32.2) v_{A2}^2 = (5.483) v_{A2}^2$$

$$PE_2 = m_B g y_{B2} = W_B y_{B2} = (150)(1) = 150 \text{ ft} \cdot \text{lb}$$

$$KE_1 + PE_1 + U_{1\to2} = KE_2 + PE_2$$

$$(0) + (175) + (0) = (5.483) v_{A2}^2 + (150)$$

$$v_{A2} = 2.135 \text{ ft/s}.$$

Using kinematics we can find the constant acceleration (which we know to be constant because all of the forces are constant):

$$v^2 = (v_0)^2 + 2a\Delta y$$

$$(2.135)^2 = (0)^2 + 2a(1)$$

$$a_A = 2.280 \text{ ft/s}^2.$$

(c) $\boxed{\vec{a}_A = 2.28 \text{ ft/s}^2 \;\downarrow}$ as expected, this matches the results from Example 20.1(c).

Figure 22.9: Newtdog as Sisyphus in Example 22.2 (© E. Diehl).

Example 22.2
Newtdog pushes a large boulder up a hill, much like Sisyphus (look it up!) in Figure 22.9. The boulder is approximately spherical, weighs one ton and is 4 ft in diameter. The hill has a 20% grade and the crest is $h = 3$ ft higher than where the boulder begins. He starts pushing the boulder from rest at the bottom of the hill at 3 ft above the ground with a constant force parallel to the slope. He stops pushing when he reaches the top and the boulder keeps rolling and reaches the bottom of the hill traveling at 15 ft/s. What constant force does Newtdog need to exert on the boulder? Ignore losses.

The boulder's mMoI when treated like a sphere is:

$$\bar{I} = \frac{2}{5}mr^2 = \frac{2}{5}\left(\frac{2000}{32.2}\right)(2)^2 = 99.38 \text{ slug} \cdot \text{ft}^2.$$

The angle of the hill's slope:

$$\beta = \tan^{-1}(0.2) = 11.31°.$$

The bolder moves up the hill a distance (along the slope):

$$\Delta s = h/\sin\beta = (3)/\sin(11.31°) = 15.30 \text{ ft.}$$

The bolder rotates up the hill an angle of:

$$\Delta\theta = \Delta s/r = (15.3)/(2) = 7.650 \text{ rad.}$$

The moment about the ground from where Newtdog is pushing:

$$M = Fy = F(3).$$

Angular speed of boulder at bottom of hill:

$$\omega = v/r = (15)/(2) = 7.500 \text{ rad/s.}$$

Treating the bottom of the hill on the left as point 1 and the bottom of the hill on the right as point 2:

$$KE_1 = 0$$

$$PE_1 = 0$$

$$U_{1 \to 2} = F \Delta s + M \Delta \theta = F(15.3) + F(3)(7.650) = F(38.25)$$

$$KE_2 = \frac{1}{2}mv_2^2 + \frac{1}{2}\bar{I}\omega_2^2 = \frac{1}{2}(2000/32.2)(15)^2 + \frac{1}{2}(99.38)(7.5)^2 = 9{,}783 \text{ ft} \cdot \text{lb}$$

$$PE_2 = 0$$

$$KE_1 + PE_1 + U_{1 \to 2} = KE_2 + PE_2$$

$$(0) + (0) + F(38.25) = (9{,}783) + (0)$$

$$F = 255.8 \text{ lb.}$$

This seems unlikely that Newtdog will be able to get the boulder up the hill. If we wanted to check, we could have him roll it to the top where it stops. In this case the end potential energy is $PE_2 = Wh = (2000)(3) = 6{,}000 \text{ ft} \cdot \text{lb}$, so the force required is 156.9 lb, which seems somewhat more likely.

Answer: $\boxed{F = 256 \text{ lb}}$.

Example 22.3
Block A ($m_A = 20$ kg) on a slope ($\theta = 60°$) in Figure 22.10 is connected to block C ($m_C = 40$ kg) by an inextensible cable and large frictionless pulley B with mass $m_B = 60$ kg and radius of $r_B = 0.25$ m. The other two frictionless pulleys have negligible mass. Both blocks are released from rest and begin to move. A spring ($k = 2000$ N/m) connected to block A is stretched $\delta = 30$ mm in the position shown. The coefficient of friction between block A and the sloped surface is $\mu = 0.25$. Find the velocity of block C when it has moved 0.5 m.

The mMoI of the pulley:

$$\bar{I}_B = \frac{1}{2}m_B r_B^2 = \frac{1}{2}(60)(0.25)^2 = 1.875 \text{ kg} \cdot \text{m}^2.$$

Figure 22.11 labels the Work-Energy datums, distances traveled and pulley datums.
The pulleys are related by the constraints of the rope length: $3s_A + s_C = l$, if we take the time derivative of these positions we get $3v_A + v_C = 0$, or $|v_A| = \frac{1}{3}|v_C|$, where we ignore the sign convention.
If we assume a time increment, we also get: $x'_A = \frac{1}{3}\Delta y_C = \frac{1}{3}(0.5) = 0.1667$ m (prime refers to distance along slope).

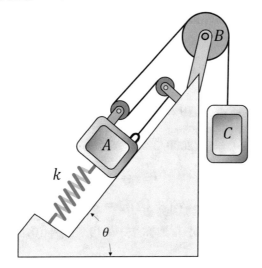

Figure 22.10: **Example** 22.3.

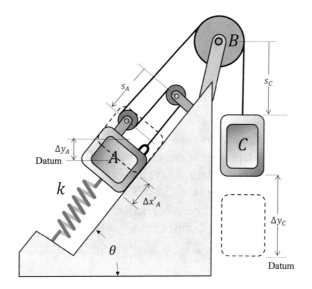

Figure 22.11: Work-energy diagram of Example 22.3.

We need to find the vertical change of block A:

$$\Delta y_A = \Delta x'_A \sin \theta = (0.1667) \sin \left(60°\right) = 0.1443 \text{ m}.$$

The rotational speed of the pulley B is related to the speed of block C by: $\omega_B = v_C / r_B$.

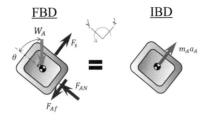

Figure 22.12: FBD/IBD pair of Example 22.3.

Figure 22.12 presents the FBD/IBD pair, although an IBD is not necessary since we only need the normal force in the y' direction which doesn't have an acceleration:

$$\nwarrow \sum F_{y'} = m_A (a_A)_{y'} = 0$$

$$- m_A g \cos \theta + F_{AN} = 0$$

$$- (20)(9.81) \cos (60°) + F_{AN} = 0$$

$$F_{AN} = 98.10 \text{ N}$$

$$F_{Af} = \mu F_{AN} = (98.10)(0.25) = 24.35 \text{ N}$$

$$KE_1 = 0$$

$$PE_1 = \frac{1}{2}k\delta_1^2 + m_C g \Delta y_C = \frac{1}{2}(2000)(0.030)^2 + (40)(9.81)(0.5) = 197.1 \text{ N} \cdot \text{m}$$

$$U_{1\to2} = -F_{Af}\Delta x_A' = -(24.35)(0.1667) = -4.088 \text{ N} \cdot \text{m} \quad \text{(negative because energy is lost).}$$

$$KE_2 = \frac{1}{2}m_A v_{A2}^2 + \frac{1}{2}\bar{I}_B \omega_B^2 + \frac{1}{2}m_C v_{C2}^2$$

$$= \frac{1}{2}(20)\left(\frac{1}{3}v_{C2}\right)^2 + \frac{1}{2}(1.875)\left(v_{C2}/(0.25)\right)^2 + \frac{1}{2}(40)v_{C2}^2 = (36.11)v_{C2}^2$$

$$PE_2 = \frac{1}{2}k\delta_2^2 + m_A g \Delta y_A$$

$$= \frac{1}{2}(2000)(0.030 + 0.1667)^2 + (20)(9.81)(0.1443) = 66.99 \text{ N} \cdot \text{m}$$

$$KE_1 + PE_1 + U_{1\to2} = KE_2 + PE_2$$

$$(0) + (197.1) - (4.088) = (36.11)v_{C2}^2 + (66.99)$$

$$v_{C2} = 1.868 \text{ m/s}.$$

Answer: $\boxed{\vec{\mathbf{v}}_{C2} = 1.87 \text{ m/s} \downarrow}$.

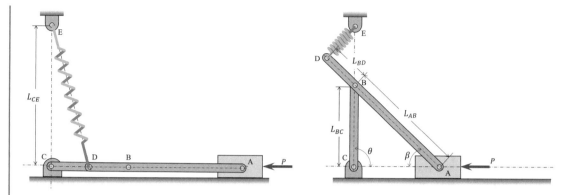

Figure 22.13: Example 22.4.

Example 22.4

The slider-crank shown in Figure 22.13 operates in the vertical plane and starts from rest in the horizontal position. The dimensions are $L_{AB} = 500$ mm, $L_{BC} = 300$ mm, $L_{BD} = 150$ mm, and $L_{CE} = 550$ mm. Each bar is 25 mm wide, 6 mm thick and is made of steel (7700 kg/m³). The spring has an unstretched length of $L_{ED} = 100$ mm and a stiffness of $k_{ED} = 10$ kN/m. A constant force of $P = 5$ kN is applied to piston A which has a mass of 5 kg. Determine the velocity of piston A when link BC is vertical.

Some preliminary geometry and mass calculations are needed. The angle of the connecting rod in the second position is:

$$\beta = \sin^{-1}\left(\frac{L_{BC}}{L_{AB}}\right) = \sin^{-1}\left(\frac{300}{500}\right) = 36.87°.$$

The initial length of spring DE is:

$$L_{ED1} = \sqrt{(L_{BC} - L_{BD})^2 + L_{CE}^2} = \sqrt{[(300) - (150)]^2 + (550)^2} = 570.1 \text{ mm}.$$

The initial stretch of spring DE is:

$$\delta_1 = L_{ED1} - L_{ED} = (570.1) - (100) = 470.1 \text{ mm}.$$

The final length of spring DE is:

$$L_{ED2} = \sqrt{(L_{BD}\cos\beta)^2 + (L_{CE} - L_{BC} - L_{BD}\sin\beta)^2}$$

$$= \sqrt{((150)\cos(36.87°))^2 + ((550) - (300) - (150)\sin(36.87°))^2} = 200 \text{ mm}.$$

The final stretch of spring DE is:

$$\delta_2 = L_{ED2} - L_{ED} = (200) - (100) = 100 \text{ mm}.$$

The vertical position of the center of mass of crank BC:

$$\Delta y_{BC} = \frac{1}{2}L_{BC} = \frac{1}{2}(300) = 150 \text{ mm}.$$

The vertical position of the center of mass of connecting rod ABD:

$$\Delta y_{ABD} = \frac{1}{2}(L_{AB} + L_{BD})\sin\beta = \frac{1}{2}((500) + (150))\sin(36.87°) = 195.0 \text{ mm}.$$

Distance piston A moves:

$$\Delta x_A = L_{AB} + L_{BC} - L_{AB}\cos\beta = (500) + (300) - (500)\cos(36.87°) = 400 \text{ mm}.$$

The mass of crank BC:

$$m_{BC} = \rho L_{BC}tw = (7700)(0.3)(0.025)(0.006) = 0.3465 \text{ kg}.$$

The mMoI of crank BC:

$$\overline{I}_{BC} = \frac{1}{12}m_{BC}L_{BC}^2 = \frac{1}{12}(0.3465)(0.3)^2 = 2.599E - 3 \text{ kg} \cdot \text{m}^2.$$

The mass of connecting rod ABD:

$$m_{ABD} = \rho L_{ABD}tw = (7700)(0.65)(0.025)(0.006) = 0.7508 \text{ kg}.$$

The mMoI of connecting rod ABD:

$$\overline{I}_{ABD} = \frac{1}{12}m_{ABD}L_{ABD}^2 = \frac{1}{12}(0.7508)(0.65)^2 = 2.643E - 2 \text{ kg} \cdot \text{m}^2.$$

The kinematics of connecting rod ABD are established most easily using the instantaneous center of rotation. Doing so reveals that at the instant shown the lines perpendicular to the motion at A and B are parallel, so there is no ICR and therefore no rotation of ABD and $v_A = v_B$. The center of mass of connecting rod ABD is also equal to this velocity. The angular speed of crank BC is found from $\omega_{BC} = v_B/L_{BC} = v_A/L_{BC}$. The speed of the center of crank BC is $v_{BC} = \omega_{BC}L_{BC}/2 = v_A/2$:

$$KE_1 = 0$$

$$PE_1 = \frac{1}{2}k\delta_1^2 = \frac{1}{2}(10{,}000)(0.4701)^2 = 1{,}105 \text{ N} \cdot \text{m}$$

$$U_{1\to2} = P\Delta x_A = (5{,}000)(0.400) = 2{,}000 \text{ N} \cdot \text{m}$$

$$KE_2 = \frac{1}{2}m_A v_{A2}^2 + \frac{1}{2}\overline{I}_{BC}\omega_{BC}^2 + \frac{1}{2}m_{BC}v_{BC}^2 + \frac{1}{2}\overline{I}_{ABD}\omega_{ABD}^2 + \frac{1}{2}m_{ABD}v_{ABD2}^2$$

$$= \frac{1}{2}(5)v_{A2}^2 + \frac{1}{2}(2.599E - 3)(v_A/(0.3))^2 + \frac{1}{2}(0.3465)(v_A/2)^2$$

$$+ \frac{1}{2}(2.643E - 2)(0)^2 + \frac{1}{2}(0.7508)(v_A)^2$$

$$= (2.933)v_{C2}^2$$

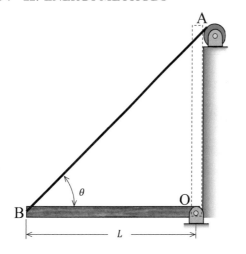
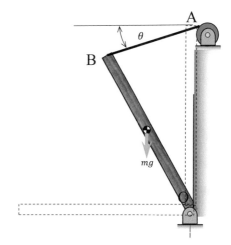

Figure 22.14: Drawbridge in Example 22.5.

$$PE_2 = \frac{1}{2}k\delta_2^2 + m_{BC}g\Delta y_{BC} + m_{ABD}g\Delta y_{ABD}$$
$$= \frac{1}{2}(10,000)(0.1)^2 + (0.3465)(9.81)(0.15) + (0.7508)(9.81)(0.195)$$
$$= 51.95 \text{ N} \cdot \text{m}$$

$$KE_1 + PE_1 + U_{1\to2} = KE_2 + PE_2$$

$$(0) + (1,105) + (2,000) = (2.933)v_{A2}^2 + (51.95)$$

$$v_{A2} = 32.26 \text{ m/s}.$$

Answer: $\boxed{\vec{v}_{A2} = 32.3 \text{ m/s} \leftarrow}$.

Example 22.5 (similar to Example 20.4)
The drawbridge in Figure 22.14 is $L_{AB} = 5$ m long and with mass $m_{AB} = 1{,}000$ kg. Starting from rest in the horizontal position, a motor-powered winch pulls a single cable to raise it. The winch drum has a 600 mm diameter, 300 kg mass, and 200 mm radius of gyration. The motor operates at a constant torque of 9 N · m. Determine the angular speed of the drawbridge when it reaches vertical and the power drawn by the motor at that instant if it is 85% efficient:

$$\overline{I}_{\text{drum}} = m_{\text{drum}}r_g^2 = (300)(0.2)^2 = 12.00 \text{ kg} \cdot \text{m}^2$$

$$\overline{I}_{OB} = \frac{1}{12}m_{OB}L_{OB}^2 = \frac{1}{12}(1000)(5)^2 = 2{,}083 \text{ kg} \cdot \text{m}^2$$

$$I_O = \overline{I}_{OB} + m_{OB}(L_{OB}/2)^2 = (2{,}083) + (1000)((5)/2)^2 = 8{,}333 \text{ kg} \cdot \text{m}^2$$

$$L_{AB} = \sqrt{L^2 + L^2} = \sqrt{2(5)^2} = 7.071 \text{ m} = r_{\text{drum}}\Delta\theta \quad \Delta\theta = \frac{L_{AB}}{r_{\text{drum}}} = \frac{(7.071)}{(0.3)} = 23.57 \text{ rad.}$$

Speed of Point B:

$$v_B = \omega_{OB} L_{OB} \qquad \omega_{\text{drum}} = v_B/r_{\text{drum}} = \omega_{OB} L_{OB}/r_{\text{drum}}$$

$$KE_1 + PE_1 + U_{1\to 2} = KE_2 + PE_2$$

$$KE_1 = 0$$

$$PE_1 = 0$$

$$U_{1\to 2} = M\Delta\theta = (90)(23.57) = 2{,}121 \text{ N} \cdot \text{m}$$

$$KE_2 = \frac{1}{2}I_O \omega_{OB2}^2 + \frac{1}{2}\overline{I}_{\text{drum}}\omega_{\text{drum}2}^2$$

$$= \frac{1}{2}(2{,}083)\omega_{OB2}^2 + \frac{1}{2}(12.00)[\omega_{OB2}(5)/(0.3)]^2 = (2{,}708)\omega_{OB2}^2$$

$$PE_2 = mgL/2 = (1000)(9.81)(5)/2 = 1{,}962 \text{ N} \cdot \text{m}$$

$$(0) + (0) + (2{,}121) = (2{,}708)\omega_2^2 + (1{,}962)$$

$$\omega_2 = 0.2423 \text{ rad/s}$$

$$\mathbb{P}_2 = M\omega_2 = (90)(0.2423) = 21.81 \frac{\text{N} \cdot \text{m}}{\text{s}} = 21.81 \text{ W.}$$

This is the delivered power. The required motor power is

$$\mathbb{P}_{motor2} = \frac{\mathbb{P}_2}{\eta_{motor}} = \frac{(21.81)}{(0.85)} = 25.66 \text{ W.}$$

Answers: $\boxed{\omega_2 = 0.242 \text{ rad/s}}$ and $\boxed{\mathbb{P}_{motor2} = 25.7 \text{ W}}$.

Book 2 - Class 23

https://www.youtube.com/watch?v=rLlW8DCNG4I

C L A S S   23

Rigid Body Impulse-Momentum Method

B.L.U.F. (Bottom Line Up Front)

- Angular Momentum of a rigid body is $\vec{\mathbf{H}}_G = \bar{I}_G \vec{\omega}$.

- Angular Impulse of a rigid body is $\overrightarrow{\mathbf{ANG\ IMP}}_{1 \to 2} = \int_1^2 \vec{\mathbf{M}}_G dt$.

- Angular Impulse-Momentum is: $\sum (\vec{\mathbf{H}}_G)_1 + \sum \int_1^2 \vec{\mathbf{M}}_G dt = \sum (\vec{\mathbf{H}}_G)_2$.

- For a rigid body not rotating about a fixed point $\sum \left(\bar{I}_G \vec{\omega} + \bar{\mathbf{r}} \times (m \vec{\mathbf{v}}_{cm}) \right)_1 +$ $\sum \int_1^2 \vec{\mathbf{M}}_O dt = \sum \left(\bar{I}_G \vec{\omega} + \bar{\mathbf{r}} \times (m \vec{\mathbf{v}}_{cm}) \right)_2$.

- Conservation of Angular Momentum about a fixed point is: $\sum (\vec{\mathbf{H}}_O)_1 = \sum (\vec{\mathbf{H}}_O)_2$ or $\bar{I}_1 \vec{\omega}_1 = \bar{I}_2 \vec{\omega}_2$ if the mMoI changes.

- Linear-Impulse Momentum is still used for the center of mass of rigid bodies.

23.1   ANGULAR MOMENTUM OF PARTICLES AND RIGID BODIES

Translational momentum (sometimes called linear momentum), as introduced in Class 6 (vol. 1) and used in Classes 10 (vol. 1) and 11 (vol. 1) for particles, is mass times velocity ($\vec{\mathbf{L}} = m \vec{\mathbf{v}}$). The tendency of an object in motion to stay in motion is translational momentum. This tendency remains true for rigid bodies with the caveat that we reference the momentum at the center of mass: $\vec{\mathbf{L}} = m \vec{\mathbf{v}}_{cm}$.

As discussed in Section 10.4 (vol. 1), a system of particles when rotating about an axis can possess angular momentum: $\vec{\mathbf{H}}_O = \vec{\mathbf{r}} \times m \vec{\mathbf{v}}$. A system of particles can also have a combined angular momentum. To simplify the following discussion, we deal with magnitudes rather than vectors, so $H_O = r m v$. Recall from rigid body kinematics in Class 14 we can relate the magnitude of the velocity about a fixed axis with $v = \omega r$. If we substitute we have: $H_O = r m \omega r$. Rearranging we find $H_O = m r^2 \omega$. We should recognize $m r^2$ as the mMoI because it's defined as $\bar{I} = \int r^2 dm$. We can think of a rigid body as a collection of masses moving together so it

Figure 23.1: Newtdog's somersaults has both translational and angular momentum (© E. Diehl).

should be clear that the angular momentum magnitude of a rigid body is $H_O = \bar{I}\,\omega$. In vector form about the center of gravity it is: $\vec{\mathbf{H}}_G = \bar{I}_G\vec{\boldsymbol{\omega}}$.

Angular momentum of a rigid body can best be thought of as a rotating object's tendency to remain rotating unless an external moment (for a duration of time) acts upon it. The direction of the angular momentum vector is the axis it's rotating about. Angular momentum can also be thought of as an object's resistance to start spinning. That is, we must add an external moment for a period of time (which we call "angular impulse") to change angular momentum.

Rigid bodies often have both translational and angular momentum such as Newtdog doing a cartwheel in Figure 23.1. His arm touching the floor will exert a force for a short duration which will help increase his translational and angular momentum to complete the cartwheel.

23.2   ANGULAR IMPULSE

An impulse, as described in Section 10.1 (vol. 1), is a force applied for a duration of time, or simply force times time: $IMP = F t$. We'll recall that it is a vector quantity that acts in the direction of the force. Similarly we have an angular impulse which is moment times time: $ANG\ IMP = M_O t$. This too is a vector quantity but the associated direction is the unit vector about which the moment is referencing. The two vector forms of the linear and angular impulses are written as integrals of the applied external loads since they might vary with time:

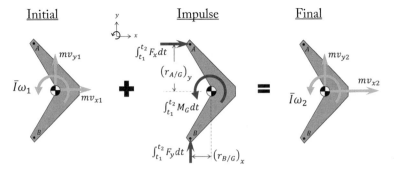

Figure 23.2: Impulse-Momentum diagram for a rigid body.

Linear impulse:

$$\overrightarrow{\textbf{IMP}}_{1\to 2} = \int_1^2 \overrightarrow{\textbf{F}}_{ext} dt.$$

Angular Impulse about center of mass:

$$\overrightarrow{\textbf{ANG IMP}}_{1\to 2} = \int_1^2 \overrightarrow{\textbf{M}}_G dt.$$

If the load is constant, or we just want to use the average load, we can write these as:

$$\overrightarrow{\textbf{IMP}}_{1\to 2} = \overrightarrow{\textbf{F}}_{ext}\Delta t.$$

Angular Impulse about center of mass:

$$\overrightarrow{\textbf{ANG IMP}}_{1\to 2} = \overrightarrow{\textbf{M}}_G \Delta t.$$

23.3 ANGULAR IMPULSE-MOMENTUM APPLIED TO RIGID BODIES

To apply Impulse-Momentum to a rigid body we use the same three stage diagram we introduced in Class 10 (vol. 1) and shown in Figure 23.2:

Before	+	External Impulse	=	After
(System Momenta)$_1$	+	(System External Impulse)$_{1\to 2}$	=	(System Momenta)$_2$

Both linear (translational) and angular momentum are applied to this diagram. The general equations we use for this are:

$$\sum \overrightarrow{\textbf{L}}_1 + \sum \int_1^2 \overrightarrow{\textbf{F}}_{ext} dt = \sum \overrightarrow{\textbf{L}}_2 \qquad \sum (\overrightarrow{\textbf{H}}_G)_1 + \sum \int_1^2 \overrightarrow{\textbf{M}}_G dt = \sum (\overrightarrow{\textbf{H}}_G)_2.$$

It's good to note the units of angular impulse and momentum in S.I. are $kg \cdot m^2 \cdot \frac{rad}{s} \Rightarrow$ $N \cdot m \cdot s$, and in U.S. units are $slug \cdot ft^2 \cdot \frac{rad}{s} \Rightarrow lb \cdot ft \cdot s$.

If the object pivots about a fixed non-CG point, we must also consider the motion of its center relative to the axis (similar to the concept of an "effective moment" in rigid body N2L), so we use the following equation instead:

$$\sum \left(\overline{I}_G \vec{\omega} + \vec{r} \times (m\vec{v}_{cm})\right)_1 + \sum \int_1^2 \vec{M}_O dt = \sum \left(\overline{I}_G \vec{\omega} + \vec{r} \times (m\vec{v}_{cm})\right)_2 .$$

When there are multiple rigid bodies, they can be treated either together as a system or treated individually. This fact can be used as a strategy to eliminate internal forces and moments that are unknown but not of interest as part of the final objective. This is because forces between bodies within a system are internal and do not contribute to the impulse because they are equal and opposite and therefore cancel. This leads us to some generalized concept equations:

$$\sum \left(\begin{array}{c} \text{System} \\ \text{Linear} \\ \text{Momentum} \end{array} \right)_1 + \sum \left(\begin{array}{c} \text{System} \\ \text{Linear Ext.} \\ \text{Impulses} \end{array} \right) = \sum \left(\begin{array}{c} \text{System} \\ \text{Linear} \\ \text{Momentum} \end{array} \right)_2$$

$$\sum \left(\begin{array}{c} \text{System} \\ \text{Angular} \\ \text{Momentum} \end{array} \right)_1 + \sum \left(\begin{array}{c} \text{System} \\ \text{Angular Ext.} \\ \text{Impulses} \end{array} \right) = \sum \left(\begin{array}{c} \text{System} \\ \text{Angular} \\ \text{Momentum} \end{array} \right)_2$$

23.4   CONSERVATION OF ANGULAR MOMENTUM

As with linear momentum, if there are no external angular impulses, angular momentum is conserved:

$$\sum (\vec{H}_O)_1 = \sum (\vec{H}_O)_2 \text{ or for simple problems with changing mMoI: } \overline{I}_1 \vec{\omega}_1 = \overline{I}_2 \vec{\omega}_2.$$

A demonstration of the conservation of angular momentum can be seen when an ice skater performs as spin with their arms outstretched and then tucks them to their body. The results are their spin rate greatly increases. This is because the angular momentum remains the same but they've reduced their mMoI, so: $\overline{I}_{large}\omega_{slow} = \overline{I}_{small}\omega_{fast}$. Newtdog performs a similar trick spinning on a swivel stool in Figure 23.3 which we'll work through in Example 23.2. Try this yourself with some dumbbells and an office chair.

It should be noted that situations arise where angular momentum is conserved even if linear momentum is not. How is that possible? If external forces are applied that align with the axis about which the angular momentum is referencing, the linear momentum will change but the angular momentum will remain the same. Also note that Angular Momentum can be conserved while energy is not conserved as with the ice skater or Newtdog spinning example. How is that possible? The act of bringing one's arms inward (or outward) is external energy (work) added

Figure 23.3: Newtdog demonstrates conservation of angular momentum on a swivel chair (© E. Diehl).

to the system via their muscles, BUT the forces exerted are internal so don't contribute to the angular momentum.

23.5   P.A.T. REVEALED WHEN APPLYING IMPULSE-MOMENTUM TO A FIXED ROTATING BODY

If an object rotates about a non-CG point without external moments, ignoring the vectors we can write:

$$\sum \left(\bar{I}\omega + rmv_{cm} \right)_1 = \sum \left(\bar{I}\omega + rmv_{cm} \right)_2 .$$

The speed of the center of mass about a fixed point is $v_{cm} = r\omega$, so substituting we find the P.A.T. again, just as we did in Sections 20.3 and 22.3:

$$\sum \left(\left(\underbrace{\bar{I} + mr^2}_{\text{P.A.T.}} \right) \omega \right)_1 = \sum \left(\left(\underbrace{\bar{I} + mr^2}_{\text{P.A.T.}} \right) \omega \right)_2$$

$$\sum \left(I_O \omega \right)_1 = \sum \left(I_O \omega \right)_2 .$$

We'll demonstrate the application of Impulse-Momentum to rigid bodies in the following examples.

Example 23.1   (This is a repeat of Examples 20.1 and 22.1)
Just like Examples 6.3, 8.2, 10.2, (vol. 1), 20.1, and 22.1, the three weights and pulley setups shown in Figure 22.4 begin at rest. The single frictionless pulley weighs $W_C = 50$ lb, has radius $r_C = 1$ ft, and a radius of gyration of $k_C = 0.75$ ft. In setup (a) the force ($P = 50$ lb) is applied to a cable attached to block A ($W_A = 75$ lb). Setup (b) has the same block A and is connected

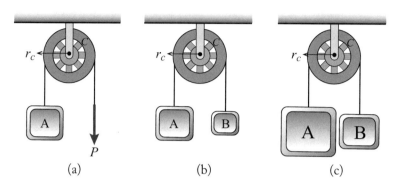

Figure 23.4: Example 23.1 mass and pulley problem. (Repeat of Figures 20.4 and 22.5.)

to block B ($W_B = 50$ lb). Setup (c) has larger blocks with the same difference between them ($W_A = 175$ lb and $W_B = 150$ lb). Determine the acceleration of block A for each setup.

The mMoI of the pulley was found using the provided radius of gyration: $I_C = 0.8734$ slug $\cdot$ ft^2.

We'll use Angular Impulse-Momentum to solve the same problem as Examples 20.1 and 22.1 which use Rigid Body N2L and Work-Energy, respectively. We again use kinematics to relate the translational velocity of the blocks and cables to the angular velocity of the pulley by $v_A = \omega_C r_C$ and $v_B = \omega_C r_C$. Just as with Example 10.2 (vol. 1), this problem asks for acceleration but Impulse-Momentum deals with velocity and time. Since the forces are constant, we can assume constant acceleration, use an assumed time increment to find the velocity after that period, and then use constant acceleration kinematics to find the acceleration. We'll use $\Delta t = 1$ s just as we did in Example 10.2 (vol. 1).

Part (a): The Impulse-Momentum diagram of part (a) is shown in Figure 23.5. Unlike Example 10.2 (vol. 1), we can include the pulley and mass in one diagram to solve this because we can eliminate the reaction forces at the pin by taking the Angular Impulse-Momentum about it:

$$\circlearrowleft \sum \left(\overline{I}_G \vec{\omega} + \vec{r} \times (m \vec{v}_{cm}) \right)_1 + \circlearrowleft \sum \vec{M}_O \Delta t = \circlearrowleft \sum \left(\overline{I}_G \vec{\omega} + \vec{r} \times (m \vec{v}_{cm}) \right)_2$$

$$(0) + W_A \Delta t\, r_C - P \Delta t\, r_C = m_A v_{A2} r_C + \overline{I}_C \omega_{C2}.$$

Replace $\omega_{C2} = \frac{v_{A2}}{r_C}$

$$(0) + (75)\,(1)\,(1) - (50)\,(1)\,(1) = \frac{(75)}{(32.2)} v_{A2}\,(1) + (0.8734)\frac{v_{A2}}{(1)}$$

$$v_{Ay2} = 7.806 \text{ ft/s}.$$

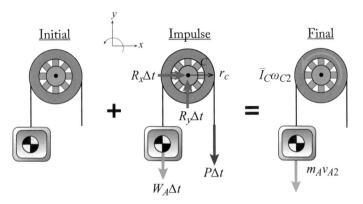

Figure 23.5: Impulse-Momentum diagram for Example 23.1(a).

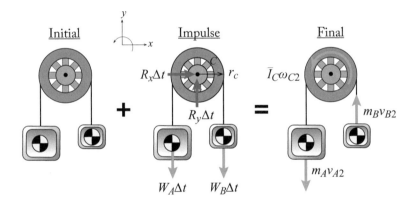

Figure 23.6: Impulse-Momentum diagram for Example 23.1(b).

Acceleration can be found from:

$$v = v_0 + at$$
$$(7.806) = (0) + a\,(1)$$
$$a = 7.806 \text{ ft/s}^2.$$

(a) $\boxed{\vec{\mathbf{a}}_A = 7.81 \text{ ft/s}^2 \downarrow}$. This matches the answers in Examples 20.1 and 22.1.

The Impulse-Momentum diagram for Part (b) is shown in Figure 23.6.

$$(0) + W_A \Delta t\, r_C - W_B \Delta t\, r_C = m_A v_{A2} r_C + m_B v_{B2} r_C + \bar{I}_C \omega_{C2}.$$

Replace $\omega_{C2} = \frac{v_{A2}}{r_C}$ and $v_{B2} = v_{A2}$

$$(0) + (75)\,(1)\,(1) - (50)\,(1)\,(1) = \frac{(75)}{(32.2)} v_{A2}\,(1) + \frac{(50)}{(32.2)} v_{A2}\,(1) + (0.8734)\frac{v_{A2}}{(1)}$$

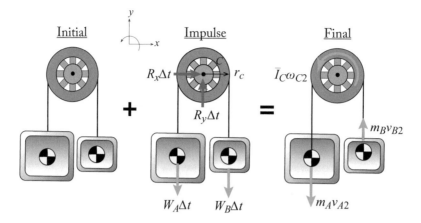

Figure 23.7: Impulse-Momentum diagram for Example 23.1(c).

$$v_{Ay2} = 5.257 \text{ ft/s}.$$

Acceleration can be found from:

$$v = v_0 + at$$

$$(5.257) = (0) + a(1)$$

$$a = 5.257 \text{ ft/s}^2.$$

(b) $\boxed{\vec{\mathbf{a}}_A = 5.26 \text{ ft/s}^2 \downarrow}$. This too matches the answers in Examples 20.1 and 22.1.

Last, the Impulse-Momentum diagram for Part (c) is shown in Figure 23.7:

$$(0) + W_A \Delta t \, r_C - W_B \Delta t \, r_C = m_A v_{A2} r_C + m_B v_{B2} r_C + \bar{I}_C \omega_{C2}.$$

Again, replace $\omega_{C2} = \frac{v_{A2}}{r_C}$ and $v_{B2} = v_{A2}$

$$(0) + (175)(1)(1) - (150)(1)(1) = \frac{(175)}{(32.2)} v_{A2}(1) + \frac{(150)}{(32.2)} v_{A2}(1) + (0.8734)\frac{v_{A2}}{(1)}$$

$$v_{Ay2} = 2.280 \text{ ft/s}.$$

Acceleration can be found from:

$$v = v_0 + at$$

$$(2.280) = (0) + a(1)$$

$$a = 2.28 \text{ ft/s}^2.$$

(c) $\boxed{\vec{\mathbf{a}}_A = 2.28 \text{ ft/s}^2 \downarrow}$. This too matches the answers in Examples 20.1 and 22.1.

Figure 23.8: Newtdog spins on a swivel stool in Example 23.2 (© E. Diehl).

We reflect one last time on the results of this problem we've solved six times. We demonstrated that each of the three kinetics methods could be used in this situation if we use kinematics as well. We noted that constant forces (and consequently constant accelerations) made this relatively easy. Not every kinetics problem can be solved using all three methods. We also noted that the difference between force and weight on cable B when comparing (a) and (b) each time resulted in less acceleration due to the resistance to motion from the added mass on B. Last, we noted the sluggishness that using larger masses and mMoI had on the results.

Example 23.2
Newtdog is spinning in his chair holding his arms outward and then pulling them inward in Figure 23.8. If Newtdog is spinning at a rate of 4 s per revolution with his arms and legs outstretched, how fast will he spin if he brings them inwards. Make some assumptions.

The initial angular velocity of Newtdog is $n_1 = \frac{(60 \text{ s/min})}{(4 \text{ s/rev})} = 15$ rpm. We don't need to convert this into radians per second but will anyway: $\omega_1 = \frac{(15 \text{ rev/min})(2\pi \text{ rad/rev})}{(60 \text{ s/min})} = 1.571$ rad/s.

We will estimate that Newtdog weighs around 70 kg (154 lbs). To find his mMoI, we'll treat his body parts as cylinders (Figure 23.9), estimate their volumes, distribute his total weight evenly, estimate the individual mMoIs and combine for the total mMoI. This is an example of modeling where we attempt to use assumptions and approximations to mimic reality so we can estimate a predicted result. We might make a much more sophisticated model using 3D solid modeling software if we desired, but that too would be an estimation.

The body is estimated to be 150 mm radius and 500 mm tall, for a volume of 0.03534 m^3.

The head is estimated to be 75 mm radius and 250 mm high, for a volume of 0.004418 m^3.

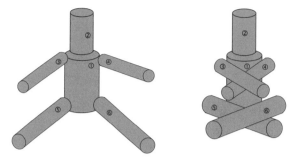

Figure 23.9: mMoI models of Newtdog in Example 23.2.

Each arm is estimated to be 50 mm radius and 500 mm long, for a volume of 0.003927 m³ each.

Each leg is estimated to be 75 mm radius and 600 mm long, for a volume of 0.01060 m³ each.

The total body volume is 0.06882 m³, for a density of $\rho = (70) / (0.06882) = 1{,}017$ kg/m³.

The mass of each body part is estimated by this density times the volume.

The head and body mMoI are treated as cylinders about their long axis using $\bar{I} = \frac{1}{2}mr^2$.

The arms and legs mMoI are treaded as cylinders about a side axis, or $\bar{I} = \frac{1}{2}m\left(3r^2 + l^2\right)$.

The center of the outstretched legs away from the axis are estimated to be 450 mm and the center of the arms 400 mm.

The center of the legs brought inward is estimated to be 225 mm and arms 200 mm.

Table 23.1 summarizes the mMoI calculation.

This is a conservation of angular momentum problem:

$$+ \circlearrowleft \sum (\vec{\mathbf{H}}o)_1 = \sum (\vec{\mathbf{H}}o)_2$$

$$\bar{I}_{y1}\omega_1 = \bar{I}_{y2}\omega_2$$

$$(6.912)\,(1.571) = (2.6774)\,\omega_2$$

$$\omega_2 = 4.055 \text{ rad/s.}$$

Table 23.1: mMoI estimation of Newtdog for Example 23.2

		r	l	V	m	I_y	d_1	$I_y + md_1^2$	d_2	$I_y + md_2^2$
		m	m	m^3	kg	kg*m^2	m	kg*m^2	m	kg*m^2
Body	1	0.15	0.5	0.03534	35.95	0.4044	0.000	0.4044	0.0000	0.4044
Head	2	0.075	0.25	0.004418	4.494	0.0126	0.000	0.0126	0.0000	0.0126
Arm 1	3	0.05	0.5	0.003927	3.994	0.0857	0.400	0.7248	0.2000	0.2455
Arm 2	4	0.05	0.5	0.003927	3.994	0.0857	0.400	0.7248	0.2000	0.2455
Leg 1	5	0.075	0.6	0.01060	10.78	0.3387	0.450	2.523	0.2250	0.8847
Leg 2	6	0.075	0.6	0.01060	10.78	0.3387	0.450	2.523	0.2250	0.8847
			V_{total} =	0.06882			I_{y1} =	6.912	I_{y2} =	2.677

We should answer with the same "units" we were given in the problem statement, which is number of seconds per revolution:

$$t = \frac{(2\pi \text{ rad/rev})}{(4.055 \text{ rad/s})} = 1.549 \text{ s/rev}.$$

Answer: $\boxed{t = 1.55 \text{ s/rev}}$.

This was a rough estimate with many assumptions to get the mMoI. Note that the swivel chair mMoI was omitted. This is an interesting problem students are encouraged to try themselves with a swivel chair. Using some small dumbbells in your outstretched hands will make the experiment even more interesting.

Example 23.3
Two 5-kg disks are attached to a 3-kg rod as shown in Figure 23.10. The two disks are both spun until they reach 10 rad/s clockwise while the rod is held in place. The rod is released and begins to spin. Eventually, because of very small friction in the pins of the disks, the disks stop spinning relative to the rod (they rotate together with the rod at the same angular speed). Determine the angular velocity when this happens:

$$\overline{I}_{ABC} = \frac{1}{12}m_{ABC}L_{ABC}^2 = \frac{1}{12}(3)(1)^2 = 0.25 \text{ kg} \cdot \text{m}^2$$

$$\overline{I}_A = \overline{I}_C = \frac{1}{2}m_A r_A^2 = \frac{1}{2}(5)(0.1)^2 = 0.025 \text{ kg} \cdot \text{m}^2.$$

When relative angular velocity is zero

$$(\omega_A)_2 = (\omega_C)_2 = (\omega_{ABC})_2 = \omega_2 \quad \text{and} \quad v_A = \omega_2 r_{A/B} \quad v_C = \omega_2 r_{C/B}.$$

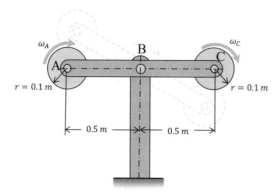

Figure 23.10: Example 23.3.

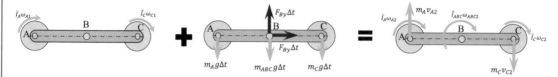

Figure 23.11: Impulse-Momentum diagram of Example 23.3.

We can see from the Impulse-Momentum diagram in Figure 23.11 that the moments balance, so this becomes a conservation of angular momentum problem. Since all of the rotations are in the clockwise direction we will treat clockwise as positive:

$$\circlearrowright \sum (\vec{\mathbf{H}}_O)_1 = \sum (\vec{\mathbf{H}}_O)_2$$

$$\bar{I}_A \omega_{A1} + \bar{I}_C \omega_{C1} = \bar{I}_{ABC} \omega_{ABC2} + \bar{I}_A \omega_{A2} + m_A v_{A2} r_{A/B} + \bar{I}_C \omega_{C2} + m_C v_C r_{C/B}$$

$$\bar{I}_A \omega_{A1} + \bar{I}_C \omega_{C1} = \bar{I}_{ABC} \omega_2 + \bar{I}_A \omega_2 + m_A r_{A/B}^2 \omega_2 + \bar{I}_C \omega_2 + m_C r_{C/B}^2 \omega_2$$

$$\underbrace{(0.025)(10) + (0.025)(10)}_{0.5}$$

$$= \underbrace{(0.25)\omega_2 + (0.025)\omega_2 + (5)(0.1)^2 \omega_2 + (0.025)\omega_2 + (5)(0.1)^2 \omega_2}_{(0.4)\omega_2}$$

$$\omega_2 = 1.250 \text{ rad/s} \circlearrowright .$$

Answer: $\boxed{\omega_2 = 1.25 \text{ rad/s} \circlearrowright}$.

Figure 23.12: Newtdog bowls in Example 23.4 (© E. Diehl).

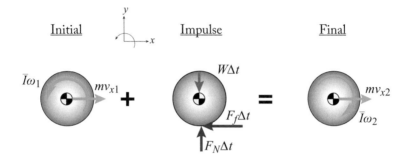

Figure 23.13: Impulse-Momentum diagram of Example 23.4.

Example 23.4

As shown in Figure 23.12, Newtdog releases his 12-lb bowling ball (mass $m_A = 5.4$ kg and diameter $d = 218$ mm) with an initial velocity of $v_1 = 4$ m/s and initial backspin of $\omega_1 = 8$ rad/s. The coefficient of kinetic friction is $\mu_k = 0.10$. Determine the time it takes for the bowling ball to stop slipping and the velocity after that happens. The mMoI of a sphere is found from $\bar{I} = \frac{2}{5}mr^2$.

The mMoI of the bowling ball is:

$$\bar{I} = \frac{2}{5}mr^2 = \frac{2}{5}(5.4)(0.109)^2 = 2.566E - 2 \text{ kg} \cdot \text{m}^2.$$

Figure 23.13 shows the Impulse-Momentum diagram of the bowling ball.

The normal force can be found from inspection to equal the weight, so $F_N = W = mg = (5.4)(9.81) = 52.97$ N. The kinetic friction is $F_f = \mu_k F_N = (0.1)(52.97) = 5.297$ N. Since the weight and normal forces are constant, the kinetic friction force is expected to be constant during the slipping process. Therefore, we can use the constant load forms of the Impulse-Momentum equations. We know that when it stops spinning the angular velocity will be $\omega_2 = $

v_{x2}/r:

$$\circlearrowleft \; \left(\overline{I}_G\vec{\omega} + \vec{r} \times (m\vec{v}_{cm})\right)_1 + \vec{M}_O\Delta t = \left(\overline{I}_G\vec{\omega} + \vec{r} \times (m\vec{v}_{cm})\right)_2$$

$$\overline{I}\omega_1 - rmv_{x1} - F_f r\Delta t = -\overline{I}\omega_2 - rmv_{x2}$$

$$(2.566E-2)(8) - (0.109)(5.4)(4) - (5.297)(0.109)\Delta t$$
$$= -(2.566E-2)v_{x2}/(0.109) - (0.109)(5.4)v_{x2}$$

$$(2.157) + (0.5774)\Delta t = (0.8240)v_{x2}. \quad \textcircled{1}$$

The bowling ball also must obey translational Impulse-Momentum principles:

$$m\vec{v}_1 + \int_{t_1}^{t_2} \vec{F}\,dt = m\vec{v}_2$$

$$mv_{x1} - F_f\Delta t = mv_{x2}$$

$$(5.4)(4) - (5.297)\Delta t = (5.4)v_{x2}$$

$$(21.60) - (5.297)\Delta t = (5.4)v_{x2}. \quad \textcircled{2}$$

Solving equations $\textcircled{1}$ and $\textcircled{2}$:

$$v_{x2} = \frac{(2.157) + (0.5774)\Delta t}{(0.8240)} = (2.618) + (0.7007)\Delta t$$

$$(21.60) - (5.297)\Delta t = (5.4)\left[(2.618) + (0.7007)\Delta t\right]$$

$$(21.60) - (5.297)\Delta t = (14.14) + (3.784)\Delta t$$

$$\Delta t = 0.8219 \text{ s}$$

$$v_{x2} = (2.157) + (0.7007)(0.8219) = 2.733 \text{ m/s}.$$

Answers: $\boxed{\Delta t = 0.822 \text{ s}}$ and $\boxed{\vec{v}_{x2} = 2.73 \text{ m/s} \rightarrow}$.

Book 2 - Class 24

https://www.youtube.com/watch?v=UQM63BnoJtU

C L A S S   24

Impact of Rigid Bodies

B.L.U.F. (Bottom Line Up Front)

- An impact on a rigid object away from its center of mass will create or exchange angular momentum.

- The "Ballistic Pendulum" problem is a common application of eccentric impact of rigid bodies.

- We still need to create a normal and tangential coordinate system, but it should be located at the point of impact. The velocities used in the CoR must be at the point of impact.

24.1   IMPACT OF RIGID BODIES

When a rigid body is struck or strikes anywhere away from the center of mass (therefore "eccentric") it creates, receives, or exchanges angular momentum. Recall from particle eccentric collisions in Class 12 (vol. 1) we created a normal and tangential local coordinate system at the impact location. We do the same with rigid body eccentric impact. The difference is that rigid bodies have shape and therefore the non-central collision will change the angular momentum.

Figure 24.1 shows a scenario where two footballs are thrown on a collision course. We can imagine what will happen: the balls will bounce off each other, each traveling upward and in opposite directions, but they will also spin away. Why? Because they've gained angular momentum due to the off-center impact.

Where they collide we create a coordinate system at the point of impact with normal and tangential axes. We break the velocities into these coordinates. In the tangential direction there is no impulse, so there is no change in the velocity in this direction. In the normal direction the final velocities are dependent on the coefficient of restitution. At this point we know the translational velocity, but how do we find the angular velocity? We apply angular impulse and momentum. We can do this about the point of impact to avoid having to include the impulse which we typically don't know. This is shown for the left football in Figure 24.2.

Applying angular Impulse-Momentum will result in the following (signs based on counter-clockwise as positive):

$$- m_A (v_{A1})_t \, x_G + m_A (v_{A1})_n \, y_G = -m_A (v_{A2})_t \, x_G - m_A (v_{A2})_n \, y_G + \overline{I}_A \omega_2.$$

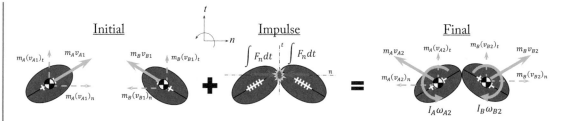

Figure 24.1: Impulse-Momentum diagram of two footballs colliding resulting in angular momentum.

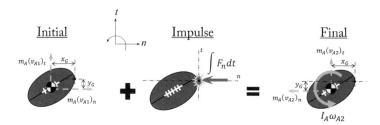

Figure 24.2: Impulse-Momentum diagram of a single football.

We note that the primary change in angular momentum comes from the change in direction of the normal velocity (it was counter-clockwise but is clockwise after the collision). Because angular momentum is conserved the ball spins counter-clockwise just as we'd guess. When we need to apply the Coefficient of Restitution (CoR), we must be sure to use the normal direction velocities *at the point of contact*, NOT at the center of mass.

24.2   ANGULAR MOMENTUM APPLIED TO BALLISTIC PENDULUMS

As discussed in Section 11.4 (vol. 1), a "ballistic pendulum" is a device used to measure the velocity of a bullet. In Example 11.4 (vol. 1) a bullet strikes a plate which we treated as if it were a particle, and we used the conservation of linear momentum in the horizontal direction for the system, the CoR, and Work-Energy to estimate the bullet speed. Since the plate was suspended by a cable, we were able to ignore the impulse forces at the connection to the top because they were entirely vertical. We also neglected the kinetic energy due to the rotation of the plate just after the bullet struck it.

If the pendulum is constructed of rigid objects, we must address the reactions at the pin in the horizontal direction. Since these are unknown (or very difficult to quantify), we can avoid them by instead applying the conservation of angular momentum about the pin.

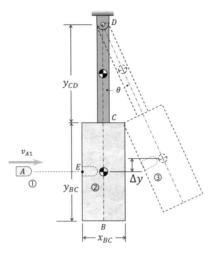

Figure 24.3: Example 24.1.

Example 24.1

A 0.25-oz bullet travels toward a 10-lb block (BC) supported a 2-lb rod (CD) and starting at rest as shown in Figure 24.3. The block is $y_{BC} = 12$ in high, $x_{BC} = 3$ in thick, and the rod is $y_{CD} = 12$ in long. The bullet embeds into the block, and the center of the block rises to a maximum height of $\Delta y = 1$ in. Determine the speed of the bullet before it hits the block.

We'll call the block in motion as state "2" and the maximum height position state "3". We'll call the center of the block where the bullet contacts Point E. The distance from the pin to E is

$$y_{DE} = y_{CD} + \frac{y_{BC}}{2} = \frac{(12)}{(12)} + \frac{(12)}{(12)2} = 1.5 \text{ ft.}$$

The mMoI of the block and rod assembly about pin D is:

$$\begin{aligned}
I_D &= \overline{I}_{BC} + m_{BC} y_{DE}^2 + \overline{I}_{CD} + m_{CD}\left(\frac{y_{CD}}{2}\right)^2 \\
&= \frac{1}{12}\left(\frac{(10)}{(32.2)}\right)\left(\left(\frac{(12)}{(12)}\right)^2 + \left(\frac{(3)}{(12)}\right)^2\right) + \left(\frac{(10)}{(32.2)}\right)(1.5)^2 \\
&\quad + \frac{1}{12}\left(\frac{(2)}{(32.2)}\right)\left(\frac{12}{12}\right)^2 + \left(\frac{(2)}{(32.2)}\right)\left(\frac{(12)}{(12)\,2}\right)^2 \\
&= 1.049 \text{ slug} \cdot \text{ft}^2.
\end{aligned}$$

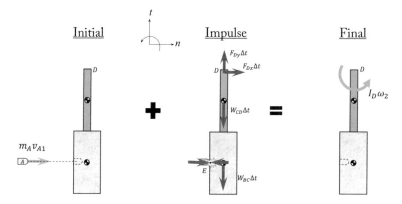

Figure 24.4: Impulse-Momentum diagram of Example 24.1.

If we include the bullet we find a very minor change:

$$I_{D\&bullet} = I_D + m_{bullet}\, y_{DE}{}^2 = (1.049) + \left(\frac{0.25/16}{32.2}\right)(1.5)^2$$

$$= (1.049) + (0.001092) = 1.050\ \text{slug} \cdot \text{ft}^2.$$

The angle at the maximum height is:

$$\theta = \cos^{-1}\left(\frac{y_{DE} - \Delta y}{y_{DE}}\right) = \cos^{-1}\left(\frac{(1.5) - (1)/12}{(1.5)}\right) = 19.19°.$$

The height change of the center of rod CD is:

$$\Delta y_{CD} = y_{CD}/2 - y_{CD}/2 \cos\theta = (12)/2 - (12)/2 \cos\left(19.19°\right) = 0.3333\ \text{in}.$$

From Conservation of Energy, using state 2 as the datum:

$$KE_2 + PE_2 = KE_3 + PE_3$$

$$\frac{1}{2}I_D\omega_2^2 + (0) = (0) + W_{BC}\Delta y_{BC} + W_{CD}\Delta y_{CD}$$

$$\frac{1}{2}(1.050)\omega_2^2 + (0) = (0) + (10)\left(\frac{(1)}{(12)}\right) + (2)\left(\frac{(0.3333)}{(12)}\right)$$

$$\omega_{BC2} = 1.301\ \text{rad/s}.$$

The Impulse-Momentum diagram shown in Figure 24.4 helps to illustrate why we cannot apply linear angular momentum to this solve for the bullet velocity: the impulsive reaction forces at the pin would need to be calculated.

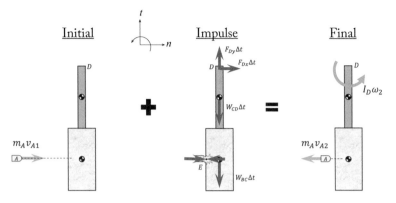

Figure 24.5: Impulse-Momentum diagram of Example 24.2.

Conservation of Angular Momentum applied about the pin allows us to avoid the impulsive reaction forces:

$$+ \circlearrowleft \sum (\vec{H}_D)_1 = \sum (\vec{H}_D)_2$$

$$m_A v_{A1} y_{DE} = I_D \omega_2$$

$$\left(\frac{0.25/16}{32.2}\right) v_{A1}(1.5) + (0) = (1.050)(1.301)$$

$$v_{A1} = 1,878 \text{ ft/s}$$

$$\frac{(1,878 \text{ ft/s})(3600 \text{ s/hr})}{(5280 \text{ ft/mi})} = 1,281 \text{ mph}$$

$$\boxed{v_{A1} = 1,280 \text{ mph}}.$$

Example 24.2
Repeat Example 24.1 but now the bullet ricochets off with a CoR of 0.4. Determine the speed of the bullet before it hits the block and its ricochet speed.

Figure 24.5 shows the updated Impulse-Momentum diagram with the bullet ricocheting off the block directly backward (even though this seems unlikely as any minor irregularity to the flight path, including the descent of the bullet due to gravity, will cause an angle to the ricochet).

We update the Conservation of Energy for the very minor change in moment of inertia due to the missing bullet:

$$KE_2 + PE_2 = KE_3 + PE_3$$

$$\frac{1}{2} I_D \omega_2^2 + (0) = (0) + W_{BC}\Delta y_{BC} + W_{CD}\Delta y_{CD}$$

$$\frac{1}{2}(1.049)\omega_2^2 + (0) = (0) + (10)\left(\frac{(1)}{(12)}\right) + (2)\left(\frac{(0.3333)}{(12)}\right)$$

$$\omega_{BC2} = 1.302 \text{ rad/s.}$$

From Conservation of Angular Momentum

$$+ \circlearrowleft \sum (\vec{\mathbf{H}}_D)_1 = \sum (\vec{\mathbf{H}}_D)_2$$

$$m_A v_{A1} y_{DE} = I_D \omega_2 - m_A v_{A2} y_{DE}$$

$$\left(\frac{0.25/16}{32.2}\right) v_{A1}(1.5) + (0) = (1.049)(1.302) - \left(\frac{0.25/16}{32.2}\right) v_{A2}(1.5)$$

$$(7.279E - 4) v_{A1} = (1.366) - (7.279E - 4) v_{A2}$$

$$v_{A1} = (1,876) - v_{A2}. \quad \textcircled{1}$$

We are left with two unknowns and need another equation to find the final answer. We use the coefficient of restitution as the second equation.

We call the point of contact on the block Point E. If we ignore the width of the block, the kinematic relationship between angular velocity and velocity at E is:

$$v_{E2} = \omega_{BC2} y_{DE} = (1.302)(1.5) = 1.953 \text{ ft/s} \rightarrow .$$

CoR using the velocities at the contact point:

$$v_{E2} - v_{A2} = e [v_{A1} - v_{E1}]$$

$$(1.953) - v_{A2} = (0.4)[v_{A1} - (0)]$$

$$(1.953) = (0.4) v_{A1} + v_{A2}. \quad \textcircled{2}$$

Combine equations $\textcircled{1}$ and $\textcircled{2}$ to find the initial and ricochet velocities. Note that the sign conventions hold true here:

$$(1.953) = (0.4)[(1,876) - v_{A2}] + v_{A2}$$

$$(1.953) = (0.4)(1,876) - (0.4) v_{A2} + v_{A2}$$

$$(1.953) - (0.4)(1,876) = (0.6) v_{A2}$$

$$v_{A2} = -1,247 \text{ ft/s.}$$

We note that the negative tells us we were wrong about our assumed direction: the bullet keeps moving to the right, although the block will still be in its way so perhaps it drops downward:

$$v_{A1} = (1,876) - (-1,247) = 3,123 \text{ ft/s} \rightarrow$$

Figure 24.6: Newtdog plays croquet in Example 24.3 (© E. Diehl).

$$\frac{(3{,}123 \text{ ft/s}) (3600 \text{ s/hr})}{(5280 \text{ ft/mi})} = 2{,}130 \text{ mph}$$

$$\boxed{v_{A1} = 2{,}130 \text{ mph}}.$$

We find a much larger initial speed than the previous example. This means it would take a much greater speed for the pendulum to swing up to the height it achieved in Example 24.1.

Example 24.3

Newtdog is playing croquet with a mallet that is 900 mm from his grip to the center of the face, has mass of 1.4 kg and has a radius of gyration about the grip of 780 mm (Figure 24.6). He raises the mallet's center of mass 150 mm off the ground and lets it swing down pivoting about his grip to strike the 0.45-kg croquet ball. What is the croquet ball's speed after being struck if the coefficient of restitution is $e = 0.8$?

We model the mallet as if it were a pendulum pivoting at his grip so the mMoI of the mallet about the grip is:

$$I_C = m_{mallet} r_{mallet}^2 = (1.4)(0.780)^2 = 0.8518 \text{ kg} \cdot \text{m}^2$$

$$KE_1 + PE_1 = KE_2 + PE_2$$

$$(0) + m_{mallet} g \Delta y = \frac{1}{2} I_C \omega_2^2 + (0)$$

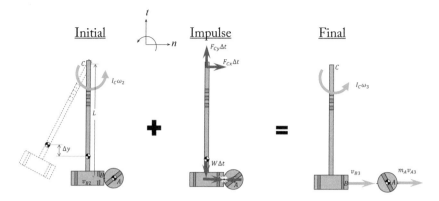

Figure 24.7: Impulse-Momentumm diagram of Example 24.3.

$$(0) + (1.4)(9.81)(0.150) = \frac{1}{2}(0.8518)\,\omega_2^2 + (0)$$

$$\omega_2 = 2.199 \text{ rad/s.}$$

The Impulse-Momentum diagram is shown in Figure 24.7.
From Conservation of Angular Momentum

$$+ \circlearrowleft \sum (\vec{\mathbf{H}}_D)_1 = \sum (\vec{\mathbf{H}}_D)_2$$

$$I_C \omega_2 = I_C \omega_3 + m_A v_{A3} L$$

$$(0.8518)(2.199) = (0.8518)\,\omega_3 + (0.45)\,v_{A3}(0.900)$$

$$(1.873) = (0.8518)\,\omega_3 + (0.4050)\,v_{A3}$$

In order to apply the coefficient of restitution we must use the velocity at the point of contact which we call Point B. We can substitute for the angular velocity after the strike:

$$\omega_3 = v_{B3}/L = v_{B3}/(0.900)$$

$$(1.873) = (0.8518)\,v_{B3}/(0.900) + (0.4050)\,v_{A3}$$

$$(1.873) = (0.9464)\,v_{B3} + (0.4050)\,v_{A3}. \quad \text{①}$$

The velocity of the mallet at the point of contact just before it strikes the croquet ball is:

$$v_{B2} = \omega_2 L = (2.199)(0.900) = 1.979 \text{ m/s.}$$

CoR

$$v_{A3} - v_{B3} = e\,[v_{B2} - v_{A2}]$$

$$v_{A3} - v_{B3} = (0.8)\,[(1.979) - (0)]$$

Figure 24.8: Newtdog bowling in Example 24.4 (© E. Diehl).

$$v_{B3} = v_{A3} - (1.583) . \quad ②$$

Combine equations ① and ②:

$$(1.873) = (0.9464) [v_{A3} - (1.583)] + (0.4050) v_{A3}$$

$$v_{A3} = 2.495 \text{ m/s}$$

$$\boxed{v_{A3} = 2.50 \text{ m/s}} .$$

We can also find the mallet's speed after impact:

$$v_{B3} = (2.495) - (1.583) = 0.9118 \text{ m/s}$$

$$\omega_3 = (0.9118)/(0.900) = 1.013 \text{ rad/s}.$$

Some interesting links with information on croquet:
https://www.croquetonline.com/
http://www.oxfordcroquet.com/

Example 24.4
Newtdog uses a 12-lb bowling ball (mass $m_A = 5.4$ kg and radius $r = 100$ mm) that rolls without slipping at initial velocity $\vec{v}_1 = 2$ m/s on a horizontal surface to squarely hit a standard bowling pin which can be treated as a uniform slender bar of mass $m_B = 1.6$ kg and length $L = 380$ mm standing on end and at rest (Figure 24.8). The coefficient of restitution between the bowling ball and pins $e = 0.7$. Determine the linear and angular velocities of both the bowling ball and the pin immediately after the impact. Neglect friction forces but assume the ball is not slipping.

We designate the ball as A and pin as B. The bowling ball mMoI is

$$\overline{I}_A = \frac{2}{5} m_A r_A^2 = \frac{2}{5} (5.4) (0.100)^2 = 2.160E - 2 \text{ kg} \cdot \text{m}^2.$$

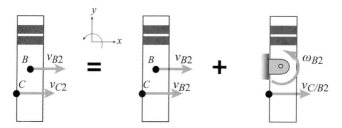

Figure 24.9: Velocity diagram of Example 24.4.

The pin mMoI treated as a slender rod is

$$\overline{I}_B = \frac{1}{12}m_B L_B^2 = \frac{1}{12}(1.6)(0.380)^2 = 1.925E - 2 \text{ kg} \cdot \text{m}^2.$$

We designate Point C as the location on the pin where the ball strikes and use it to apply the coefficient of restitution. Remember it is NOT the pin's center of mass:

$$(v_{C2} - v_{A2}) = e(v_{A1} - v_{C1}) = (0.7)((2) - (0))$$

$$v_{C2} = v_{A2} + (1.400). \quad \text{①}$$

To relate Points B and C on the pin, we use the velocity diagram of Figure 24.9.

$$v_{C2} = v_{B2} + v_{C/B2} = v_{B2} + \omega_{B2}r_{C/B} = v_{B2} + \omega_{B2}\left(\frac{L_B}{2} - r_B\right)$$

$$= v_{B2} + \omega_{B2}\left(\frac{(0.38)}{2} - (0.1)\right)$$

$$v_{C2} = v_{B2} + (0.045)\,\omega_{B2}. \quad \text{②}$$

We create an Impulse-Momentum diagram in Figure 24.10 to use for both linear and angular impulse-momentum.

System Momenta₁ + System External Impulse₁→₂ = System Momenta₂

Linear Momentum:

$$(\vec{\mathbf{L}})_1 + \left(\overrightarrow{\mathbf{IMP}}\right)_{1\to2} = (\vec{\mathbf{L}})_2$$

$$\sum m\vec{\mathbf{v}}_1 + \int_{t_1}^{t_2} \vec{\mathbf{F}}\,dt = \sum m\vec{\mathbf{v}}_2.$$

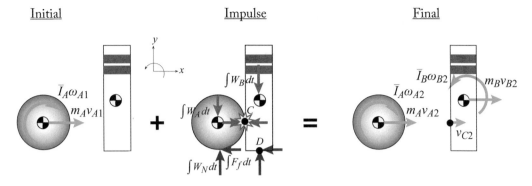

Figure 24.10: Impulse-Momentum diagram of Example 24.4.

We've drawn in the friction impulses but were told to neglect them. Therefore, we can use conservation of linear momentum.

x-dir:

$$m_A v_{A1} + (0) = m_A v_{A2} + m_B v_{B2} \quad \text{(assume ball keeps moving in positive direction)}$$

$$(5.4)(2) + (0) = (5.4) v_{A2} + (1.6) v_{B2}$$

$$(10.80) = (5.4) v_{A2} + (1.6) v_{B2}$$

$$v_{B2} = (6.750) - (3.375) v_{A2}. \quad \textcircled{3}$$

Angular Momentum about D:

$$\left(\vec{\mathbf{H}}_D\right)_1 + \left(\overrightarrow{\mathbf{Ang\ IMP}}_D\right)_{1 \to 2} = \left(\vec{\mathbf{H}}_D\right)_2$$

$$\circlearrowleft \sum \overline{I}\vec{\omega}_1 + \sum m\vec{\mathbf{v}}_1 \times \vec{\mathbf{r}}_{G/C} + \int_{t_1}^{t_2} \vec{\mathbf{M}}_C dt = \sum \overline{I}\vec{\omega}_2 + \sum m\vec{\mathbf{v}}_2 \times \vec{\mathbf{r}}_{G/C}.$$

We've selected a Point "D" where the weight, normal force and friction force of pin B and friction force (which we were told to neglect anyway) of the ball won't factor into the moment. The weight of the ball and normal force of the ball should be equal and cancel. Therefore, with this system and arrangement we can use conservation of angular momentum. We'll substitute $\omega_A = v_A / r_A$ while doing this calculation:

$$-\overline{I}_A \omega_{A1} - m_A v_{A1} r_A + (0) = \overline{I}_B \omega_{B2} - m_B v_{B2} r_{B/D} - \overline{I}_A \omega_{A2} - m_A v_{A2} r_A$$

$$-\overline{I}_A v_{A1} / r_A - m_A v_{A1} r_A + (0) = \overline{I}_B \omega_{B2} - m_B v_{B2} r_{B/D} - \overline{I}_A v_{A2} / r_A - m_A v_{A2} r_A$$

$$-(2.160E - 2)(2)/(0.1) - (5.4)(2)(0.1) + (0)$$

$$= (1.925E - 2)\,\omega_{B2} - (1.6)\,v_{B2}\left(\frac{0.38}{2}\right) - (2.160E - 2)\,v_{A2}/(0.1) - (5.4)\,v_{A2}(0.1)$$

$$- (1.512) = -(0.7560)\,v_{A2} - (0.3040)\,v_{B2} + (1.925E - 2)\,\omega_{B2}.\quad \text{④}$$

Summarizing the four equations and four unknowns:

$$v_{C2} = v_{A2} + (1.400)\quad \text{①}$$

$$v_{C2} = v_{B2} + (0.045)\,\omega_{B2}\quad \text{②}$$

$$v_{B2} = (6.750) - (3.375)\,v_{A2}\quad \text{③}$$

$$- (1.512) = -(0.7560)\,v_{A2} - (0.3040)\,v_{B2} + (1.925E - 2)\,\omega_{B2}.\quad \text{④}$$

Combine ① and ②

$$v_{A2} + (1.400) = v_{B2} + (0.045)\,\omega_{B2}\qquad \omega_{B2} = (22.22)\,v_{A2} - (22.22)\,v_{B2} + (31.11).\quad \text{⑤}$$

Combine ③ and ⑤

$$\omega_{B2} = (22.22)\,v_{A2} - (22.22)\,[(6.750) - (3.375)\,v_{A2}] + (31.11)$$

$$\omega_{B2} = (97.21)\,v_{A2} + (-118.9).\quad \text{⑥}$$

Combine ④, ⑤, and ⑥

$$- (1.512) = -(0.7560)\,v_{A2} - (0.3040)\,[(6.750) - (3.375)\,v_{A2}]$$
$$+ (1.925E - 2)\,[(97.21)\,v_{A2} + (-118.9)]$$

$$v_{A2} = 1.321 \text{ m/s} \;\rightarrow\; .$$

From ⑥

$$\omega_{B2} = (97.21)\,(1.321) + (-118.9) = 9.501 \text{ rad/s} \;\circlearrowleft\; .$$

From ③

$$v_{B2} = (6.750) - (3.375)\,(1.321) = 2.282 \text{ m/s} \;\longrightarrow\; .$$

The final angular velocity of the ball (assuming no slipping) is:

$$\omega_{A2} = v_{A2}/r_A = (1.321)/(0.1) = 13.21 \text{ rad/s} \;\circlearrowleft\; .$$

Answers:

$$\boxed{\vec{v}_A = 1.32 \text{ m/s} \;\longrightarrow\;}\qquad \boxed{\vec{\omega}_A = 13.2 \text{ rad/s} \;\circlearrowleft\;}$$

$$\boxed{\vec{v}_B = 2.82 \text{ m/s} \;\longrightarrow\;}\qquad \boxed{\vec{\omega}_B = 9.50 \text{ rad/s} \;\circlearrowleft\;}.$$

Figure 24.11: Newtdog and Wormy out.

We can picture the results, the ball keeps moving but slows a bit and the pin flies faster and spins the opposite direction. If the pin had not been treated as a slender rod, but instead had a sloping profile as in Figure 24.8, it would have had a vertical velocity component. These results make sense and this problem illustrates the principles of rigid body oblique impact. We note how we had to use kinematics and some strategy to get to a solution in this long and involved problem.

24.3 CONCLUSION

In conclusion: Dynamics is hard. The only way to get better at solving Dynamics problems is to practice solving Dynamics problems. Start with easy examples, cover up the answers, solve them all the way through to an answer, and then check your results. Keep doing this and push yourself and you will not only get better at solving Dynamics problems, you'll get better at solving other engineering problems. Following this routine is the secret to success in engineering courses.

In Figure 24.11, Newtdog and Wormy say "So long," "Rock on," and "Go chill on a beach somewhere" after finals week.

APPENDIX B

Rigid Body Dynamics Sample Exam Problems

This appendix provides sample problems for Exam 3 (Rigid Body Kinematics) and the Final Exam (Rigid Body Kinetics) corresponding to the course sechdule in Table 0.1. These are included so students can practice for the exams with problems of the approximate level of difficulty they might expect. A typical 75-min exam might include three to four of these problems, so students are encouraged to time themselves when taking them and attempt to stay under approximately 20 min each. The final exam is often longer and comprehensive with emphasis on the last topics covered (Rigid Body Kinetics), so students should prepare accordingly.

B.1 RIGID BODY KINEMATICS

These problems are covered in Classes 13–18.

B.1.1 RIGID BODY ANGULAR KINEMATICS

At the instant shown, pulley A is rotating $\omega_A = 25$ rad/s counter-clockwise and *decelerating* at a constant $\alpha_A = 12$ rad/s^2. The dimensions of the pulleys are: $r_A = 2.5$ in, $r_B = 3.5$ in, and $r_C = 1.5$ in. See Figure B.1. Determine the magnitude of the acceleration of Point B after pulley A has rotated four revolutions.

B.1.2 RIGID BODY VELOCITY ANALYSIS

A 3-ft-diameter wheel rolls to the right at $v_G = 1.25$ ft/s without slipping. Point A on the wheel is located $\theta = 30°$ clockwise from horizontal at the instant shown. See Figure B.2.
Determine:

(a) The angular speed, ω, in rad/s.

(b) The velocity of Point A in ft/s, in terms of $\hat{\imath}$ and $\hat{\jmath}$.

(c) The angular acceleration magnitude, α, in rad/s^2 if the wheel started from rest with Point A at the bottom, has not yet made a full rotation, and is accelerating at a constant rate.

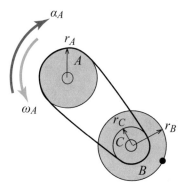

Figure B.1: Example practice problem B.1.1.

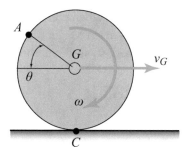

Figure B.2: Example practice problem B.1.2.

B.1.3   INSTANTANEOUS CENTER OF ROTATION 1

Link AB of the illustrated system is $L_{AB} = 0.75$ m at $\beta = 30°$ with respect to horizontal. Link BC is $L_{BC} = 1.00$ m and horizontal. Link CD is $L_{CD} = 0.5$ m at $\theta = 30°$ with respect to vertical and is rotating at $\vec{\omega}_{CD} = -4\hat{k}$ rad/s as shown. See Figure B.3. Determine the speed (velocity magnitude) of Point B using the instantaneous center of rotation method of velocity analysis.

B.1.4   INSTANTANEOUS CENTER OF ROTATION 2

Link AB of the illustrated system is $L_{AB} = 0.75$ m at $\theta = 45°$ with respect to horizontal. Link BC is $L_{BC} = 1.00$ m and horizontal. Link CD is $L_{CD} = 0.5$ m at $\beta = 60°$ with respect to horizontal and is rotating at $\vec{\omega}_{CD} = -4\hat{k}$ rad/s as shown. See Figure B.4. Determine the speed (velocity magnitude) of Point B using the instantaneous center of rotation method of velocity analysis.

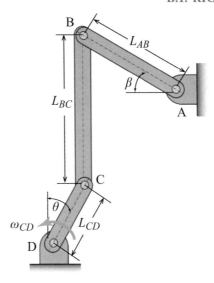

Figure B.3: Example practice problem B.1.3.

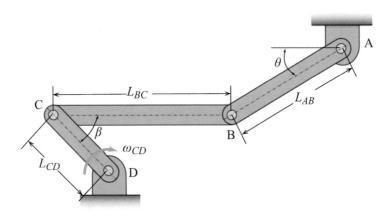

Figure B.4: Example practice problem B.1.4.

B.1.5 INSTANTANEOUS CENTER OF ROTATION 3

The linkage shown has dimensions $L_{AB} = L_{CD} = 1$ m, $L_{BC} = 3$ m, $L_P = 2$ m (the leg with P is centered and perpendicular to BC). At the instant shown, $\theta_A = \theta_D = 60°$ and $\omega_{AB} = 3.5$ rad/s clockwise. See Figure B.5. Determine the velocity of Point P.

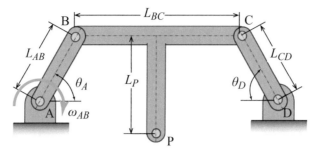

Figure B.5: Example practice problem B.1.5.

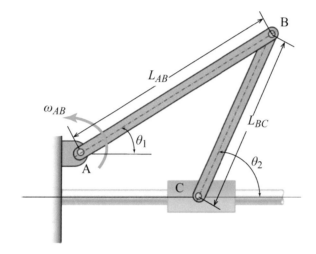

Figure B.6: Example practice problem B.1.6.

B.1.6   ACCELERATION ANALYSIS 1

Link AB is $L_{AB} = 2$ m long, positioned at $\theta_1 = 30°$ from horizontal, and connected to link BC which is $L_{BC} = 1.5$ m long and positioned $\theta_2 = 65°$ from horizontal. At the instant shown Link AB is rotating with a constant angular speed of $\omega_{AB} = 5$ rad/s counter-clockwise. See Figure B.6. Find the velocity (in m/s) and acceleration (in m/s^2) of Point C, including direction. You are required to use the instantaneous center of rotation for the velocity part of this problem.

B.1.7   ACCELERATION ANALYSIS 2

Link AB is $L_{AB} = 2$ ft long, positioned at $\theta_1 = 30°$ from vertical, and connected to link BC which is $L_{BC} = 1.5$ ft long and positioned $\theta_2 = 65°$ from vertical. At the instant shown Link AB is rotating with a constant angular speed of $\omega_{AB} = 5$ rad/s counter-clockwise. See Figure B.7. Find the velocity (in ft/s) and acceleration (in ft/s^2) of Point C, including direction. You are required to use the instantaneous center of rotation for the velocity part of this problem. What is the acceleration if link AB is also decelerating at $\alpha_{AB} = 10$ rad/s^2?

B.1.8   ACCELERATION ANALYSIS 3

Collar D is moving to the right with velocity $v_D = 4$ m/s and acceleration $a_D = 2$ m/s^2. Link BD is $L_{BD} = 500$ mm long and is $\theta = 125°$ at the instant shown. Disk AB has a radius $r_{AB} = 150$ mm. A velocity analysis shows that $\omega_{BD} = 9.766$ rad/s and $\omega_{AB} = 18.67$ rad/s, both counterclockwise. See Figure B.8. Determine the angular acceleration of disk AB.

B.2   FINAL EXAM

These problems are covered in Classes 19–24.

B.2.1   MASS MOMENT OF INERTIA

The $w = 4$ in wide by $h = 4$ in high by $t = 1$ in thick block has a $d = 1$ in Ø diameter hole which is centered horizontally ($x_h = 2$ in) and positioned $y_h = 3$ in from the bottom edge. The density is $\rho = 2$ slugs/in^3. See Figure B.9. Find the following:

(a) Total mass (slugs).

(b) Centroid position (x, y, z).

(c) mMoI about the z-axis (slugs-in^2).

B.2.2   RIGID BODY N2L 1

A $L = 3$ m board weighs $m = 100$ kg and begins at rest on a frictionless horizontal surface. A $F = 500$ N force is applied in the z-direction at $x = 0.75$ m from the end. See Figure B.10. Determine acceleration (in m/s^2) of the Point B. You are required to draw an appropriately labeled FBD/IBD pair as part of your answer.

B.2.3   RIGID BODY N2L 2

The uniform beam weighs $W_{AB} = 200$ lb, is $L_{AB} = 12$ ft long, and starts from rested when it is lifted by force A of $F_A = 250$ lb at an angle of $\theta = 65°$ and by force B of $F_B = 150$ lb straight upward. See Figure B.11. Determine the magnitude and direction of acceleration of Point A.

B.2.4   RIGID BODY WORK-ENERGY 1

Block A ($m_A = 20$ kg) on a slope ($\theta = 20°$) is connected to block C ($m_C = 40$ kg) by an inextensible cable and frictionless pulley with mass $m_B = 60$ kg and radius $r_B = 0.25$ m (treat like a disk). Both are released from rest and begin to move. A spring ($k = 2000$ N/m) connected

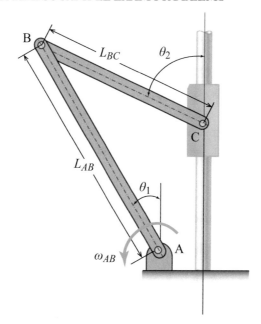

Figure B.7: **Example** practice problem B.1.7.

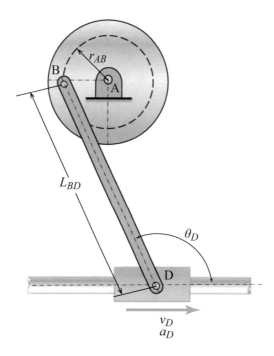

Figure B.8: **Example** practice problem B.1.8.

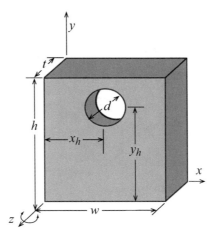

Figure B.9: Example practice problem B.2.1.

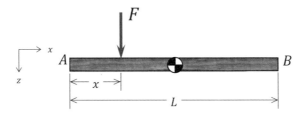

Figure B.10: Example practice problem B.2.2.

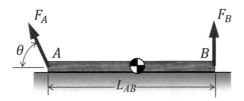

Figure B.11: Example practice problem B.2.3.

to block A is stretched $\delta = 0.02$ m in the position shown. The coefficient of friction between block A and the sloped surface is $\mu = 0.2$. See Figure B.12. Find the tension (in Newtons) of the cable between mass A and pulley B at the instant shown.

B.2.5   RIGID BODY WORK-ENERGY 2

A spherical stone with a mass of $m = 1,000$ kg and radius $r = 0.25$ m starts from rest at the top of a hill with elevation $y_1 = 100$ m. It gets nudged and begins to roll (without slipping) down

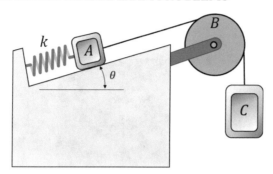

Figure B.12: Example practice problem B.2.4.

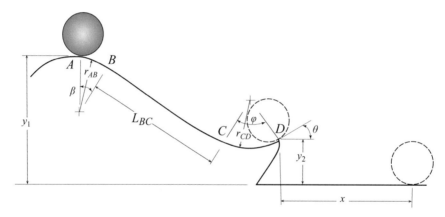

Figure B.13: Example practice problem B.2.5.

the hill. There is a cliff near the bottom of the hill where $y_2 = 10$ m that acts like a ramp with a launch angle of $\theta = 30°$. Other dimensions include: radii at the stop and of the hill and bottom near the ramp of $r_{AB} = r_{CD} = 10$ m, straight distance of slope of $L_{AB} = 120$ m, and arc angles of $\beta = 30°$ and $\varphi = 60°$. See Figure B.13. How far from the end of the ramp (x, in meters) will the stone travel before hitting the flat ground? What is the distance the stone would bounce after it lands the first time if the coefficient of restitution is $e = 0.05$?

B.2.6 RIGID BODY IMPACT 1

The pinball flipper shown ($m_{BC} = 0.25$ kg, $L_{BC} = 100$ mm, $w_{BC} = 10$ mm) operates in the horizontal plane and swings freely with a constant angular velocity $\omega_{BC} = 25$ rad/s when it strikes the pinball ($m_A = 0.1$ kg) which is moving toward it with a constant speed of $v_A = 0.1$ m/s. The coefficient of restitution is $e = 0.9$. See Figure B.14. Find the speed of the pinball (in m/s) after the paddle strikes it.

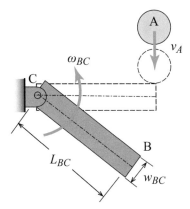

Figure B.14: Example practice problem B.2.6.

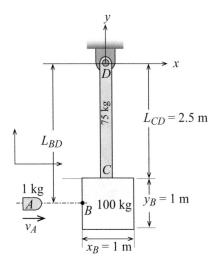

Figure B.15: Example practice problem B.2.7.

B.2.7   RIGID BODY IMPACT 2

The ballistic pendulum shown has the following parameters: $m_A = 1$ kg, $m_B = 100$ kg, $m_{CD} = 75$ kg, $L_{BD} = 3$ m, $L_{CD} = 2.5$ m, $x_B = 1$ m, and $y_B = 1$ m. The bullet embeds into the block, therefore $e = 0$. See Figure B.15. If the pendulum stops when $\theta = 15°$, what was the bullet's speed before impact? How much energy dissipates due to the collision? What if the pendulum stops at the same spot, but the bullet bounced off $(e = 1)$?

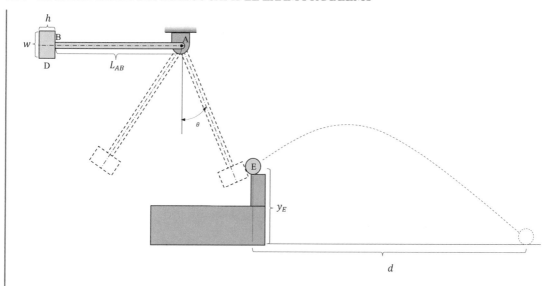

Figure B.16: Example practice problem B.2.8.

B.2.8   RIGID BODY IMPACT 3

A hammer attached to a frictionless hinge is released from horizontal and strikes ball E when it is $\theta = 30°$ past vertical. The hammer's slender handle is $L_{AB} = 1.5$ m long with mass $m_{AB} = 2$ kg, and its block head is $w = 0.25$ m wide, $h = 0.1$ m tall, and of mass $m_D = 6$ kg. The hammer head strikes the ball at the center of the face off of a frictionless surface. The ball is $d_E = 75$ mm diameter, $m_E = 0.5$ kg mass, and its center is $y_E = 1$ m from the ground. The coefficient of restitution between the ball and hammer head is $e = 0.9$. See Figure B.16.

Determine:

(a) The distance d the ball travels after impact.

(b) The maximum angle (θ measured from vertical axis) the hammer reaches after impact.

B.3   ANSWERS TO SAMPLE EXAM PROBLEMS

Table B.1 presents the answers to the sample exam problems. It is almost certain that most students will be disappointed that fully worked out solutions are not provided. I encourage you to do the difficult work of solving the problems and persevering until you find where you went astray (that is, if you don't get the same results the first time you attempt). This extra effort will pay off in your exam preparation as finding your own mistakes helps to avoid making them again.

Table B.1: Answers to sample exam problems

B.1.1 - Rigid Body Angular Kinematics	$|a_B| = 223$ in/s^2
B.1.2 – Rigid Body Velocity Analysis	$\omega = 0.833$ rad/s, $\vec{\mathbf{v}}_A = (1.88)\,\hat{\mathbf{i}} + (1.08)\,\hat{\mathbf{j}}$ m/s, $\alpha = 0.166$ rad/s^2
B.1.3 – Instantaneous Center of Rotation 1	$v_B = 1.16$ m/s
B.1.4 – Instantaneous Center of Rotation 2	$v_B = 2.45$ m/s
B.1.5 – Instantaneous Center of Rotation 3	$v_P = 5.36$ m/s $\rightarrow$
B.1.6 – Acceleration Analysis 1	$\vec{\mathbf{v}}_C = 13.6$ m/s $\rightarrow$, $\vec{\mathbf{a}}_C = 566$ m/s^2 $\rightarrow$
B.1.7 – Acceleration Analysis 2	$\vec{\mathbf{v}}_C = 6.79$ ft/s$\uparrow$, $\vec{\mathbf{a}}_C = 283$ ft/s^2 $\uparrow$, $\vec{\mathbf{a}}_C = 269$ ft/s^2 $\uparrow$
B.1.8 – Acceleration Analysis 3	$\alpha_{AB} = 153$ rad/s^2 $\circlearrowleft$
B.2.1 – Mass Moment of Inertia	$m = 30.4$ *slugs*, $\bar{x} = 2.00$ in, $\bar{y} = 1.95$ in, $\bar{z} = 0.500$ in, $I_z = 320$ *slug* $\cdot$ in^2
B.2.2 – Rigid Body N2L 1	$\vec{\mathbf{a}}_B = -2.50\,\hat{\mathbf{k}}$ m/s^2
B.2.3 – Rigid Body N2L 2	$|\vec{\mathbf{a}}_A| = 67.6$ ft/s^2, $\theta = 75.4°$
B.2.4 – Rigid Body Work-Energy 1	$F_T = 199$ *N*
B.2.5 – Rigid Body Work-Energy 2	$x = 152$ m, $\Delta x_{2-3} = 8.71$ m
B.2.6 – Rigid Body Impact 1	$\vec{\mathbf{v}}_A' = 3.76$ m/s $\uparrow$, $\eta = 83.2\%$
B.2.7 – Rigid Body Impact 2	$v_{A1} = 179$ m/s, $E_{lost} = 15{,}800$ Nm, $v_{A1} = 177$ m/s
B.2.8 – Rigid Body Impact 3	$d = 9.07$ m, $\theta = 49.4°$

Author's Biography

EDWARD DIEHL

Dr. Edward Diehl obtained his doctoral degree in Mechanical Engineering from the University of Connecticut in December 2016. He is currently an Assistant Professor at the University of Hartford in the Mechanical Engineering Department. Prior to joining UHartford, he was a lecturer (2009–2017) at the United States Coast Guard Academy in both the Mechanical Engineering section and Naval Architecture and Marine Engineering section. He worked as a Principal Engineer (2006–2009, 1996–2000, and 1992–1995) for Seaworthy Systems, Inc., self-employed (2000–2006), and an analyst (1995–1996) for General Dynamics/Electric Boat. He is a registered Professional Engineer in Connecticut. He obtained a Master of Science in Mechanical Engineering from Rensselaer at Hartford in 1996. He is a proud graduate of the United States Merchant Marine Academy at Kings Point, class of 1992, with a Bachelor of Science degree in Marine Engineering Systems. His research interests include solid mechanics pedagogy, gear vibration and fault modeling, and mechanism design.

Printed in the United States
by Baker & Taylor Publisher Services